GUIDELINES FOR FORENSIC ENGINEERING PRACTICE

SECOND EDITION

PREPARED BY
Forensic Practices Committee
Technical Council on Forensic Engineering of
the American Society of Civil Engineers

EDITED BY
Joshua B. Kardon, Ph.D., S.E., F.ASCE

ASCE

Published by the American Society of Civil Engineers

Cataloging-in-Publication Data on file with the Library of Congress.

Published by American Society of Civil Engineers
1801 Alexander Bell Drive
Reston, Virginia, 20191-4400
www.asce.org/pubs

Front cover photograph of the I35-W Bridge in Minneapolis, Minnesota, courtesy of John Weeks; all other front cover photographs courtesy of Joshua B. Kardon; back cover photograph of the I35-W Bridge collapse courtesy of Tim Davis.

ISBN 978-0-7844-1246-6 (paper)
ISBN 978-0-7844-7696-3 (PDF e-book)
Manufactured in the United States of America.

Contents

Chapter Authors, Contributors, and Reviewers

Deepak Ahuja, P.E., M.ASCE
Behnam Arya, Ph.D., P.E., M.ASCE
James Cohen, P.E., M.ASCE
Anthony Dolhon, P.E., M.ASCE
Lawrence Dombrowsk, P.E., M.ASCE
Derrick Hancock, P.E., M.ASCE
Joshua B. Kardon, Ph.D., P.E., S.E., F.ASCE
Kevin Lee McCoy, P.E., M. ASCE
Leonard J. Morse-Fortier, Ph.D., P.E., M.ASCE
Erik L. Nelson, Ph.D., P.E., M.ASCE
Irvine Lee Pack, E.I., A.M.ASCE
Robert T. Ratay, Ph.D., P.E., F.ASCE
Clemens Rossell, P.E., M.ASCE
Lloyd Sonenthal, P.E., M.ASCE

Forensic Practices Committee

Deepak Ahuja, P.E., M.ASCE, Chair
Robert B. Anderson, P.E., L.S., M.ASCE
Behnam Arya, Ph.D., P.E., M.ASCE
Steven C. Batterman, Ph.D., M.ASCE
Dan M. Bell, P.E., M.ASCE
Keith E. Brandau, P.E., M.ASCE
David M. Brani, P.E., M.ASCE
John T. Bryant, P.E., M.ASCE
Ronald J. Carrington, P.E., M.ASCE
James Cohen, P.E., M.ASCE
Edward M. Deegan, Jr., P.E., M.ASCE
Anthony M. Dolhon, P.E., M.ASCE
Lawrence Dombrowski Jr., P.E., M.ASCE
Michael J. Drerup, P.E., M.ASCE
Warren R. French, P.E., M.ASCE
Derrick S. Hancock, P.E., M.ASCE
Steven C. Helfrich, P.E., M.ASCE
James W. Jordan, P.E., M.ASCE
Harvey A. Kagan, P.E., F.ASCE
Joshua B. Kardon, Ph.D., S.E., F.ASCE
Mark E. Kubena, P.E., M.ASCE
Michael P. Lester, P.E., M. ASCE
Irvine W. Lee Pack, E.I., A.M.ASCE
Gary Lewis, D.WRE, Ph.D., P.E., F.ASCE
Kevin Lee McCoy, P.E., M.ASCE
John R. Messineo, Jr., P.E., M.ASCE
Andrew T. Metzger, P.E., M.ASCE
Leonard J. Morse-Fortier, Ph.D., P.E., M.ASCE
Michael J. Osburn, P.E., M.ASCE
Theodore G. Padgett, Jr., M.ASCE
Gregory P. Parsons, P.E., M.ASCE
David B. Peraza, P.E., M.ASCE
Robert T. Ratay, Ph.D., P.E., F.ASCE
Clemens J. Rossell, P.E., M.ASCE
Lloyd M. Sonenthal, P.E., M.ASCE
David Sykora, Ph.D., P.E., G.E., M.ASCE
Kassim M. Tarhini, P.E., M.ASCE
Glenn G. Thater, P.E, M.ASCE

Preface

Engineering investigation of buildings, bridges, and other constructed facilities that fail or do not perform as intended, rendering opinions as to the causes of failure or underperformance, and giving testimony in judicial proceedings are fields of professional practice often referred to as forensic engineering. Forensic engineering practices also are applied in circumstances involving defects, failures, and accidents involving manufactured products, consumer products, machinery, and vehicles of all types; however this publication principally addresses forensic engineering for the civil built environment.

The Technical Council on Forensic Engineering (TCFE) of the American Society of Civil Engineers (ASCE) was established in 1985, and its Forensic Practices Committee (FPC) was charged with developing guidelines for forensic engineering practice. Work on the document was started in 1994 and resulted, in 2003, in the publication of the First Edition of the Committee Report, *Guidelines for Forensic Engineering Practice.*

Design codes and standards, construction safety regulations, tools of investigation and analysis, and dispute resolution rules and procedures have evolved since 2003, when the first edition of the *Guidelines* was published. More importantly, forensic engineering has matured, becoming a more accepted, organized, and active field of practice. In recognition of these changes, the FPC reviewed the *Guidelines* and concluded that an update and revision were in order. The result of that review, update, and revision is this 2012 publication of the *Guidelines for Forensic Engineering Practice, Second Edition.*

Acknowledgments

Gary L. Lewis served as the editor for the first edition of the *Guidelines for Forensic Engineering Practice*, published by ASCE in 2003. Then and now, the *Guidelines* represent the work of the Forensic Practice Committee members and others who authored each chapter of each edition. All interested members of the Technical Council of Forensic Engineering (TCFE) reviewed these *Guidelines*. The contents of this Second Edition represent TCFE's current perspective on all aspects of these *Guidelines*, and each chapter differs from its predecessor. We owe a great debt to the committee members and authors who contributed to that earlier work, and especially to Gary Lewis for his oversight as editor of the First Edition, and for his persistence in shepherding the First Edition to print.

The editor, authors, contributors, and reviewers of the *Guidelines* wish to express their sincere thanks to all the exceptional professionals who have given selflessly over an extended period of time toward this publication. We wish to express our appreciation for the support of the TCFE Executive Committee, and of the ASCE staff. Thank you.

A NOTE ABOUT THE COVER PHOTOGRAPHS

The photographs of the I35-W Bridge (Bridge 9340) in Minneapolis, Minnesota, on the cover of this edition of the *Guidelines for Forensic Engineering Practice* have been used with the permission of the photographers. The photograph on the front cover depicts the bridge prior to its collapse on August 1, 2007, and was taken by John Weeks; the photograph on the back cover shows a portion of the collapsed bridge on that day and was taken by Tim Davis.

These photographs were selected because forensic engineering practices were applied by consultants to the State of Minnesota before the bridge collapsed to aid the State in defining and scheduling maintenance and repair work, and again after the collapse to aid many parties in their understanding of the causes and sequence of the failure.

Although the photographs both depict a bridge, the *Guidelines for Forensic Engineering Practice* is intended to address forensic practice relating to other facilities in addition to bridges, such as buildings. The photographs also are intended to provide a clear, graphic example of the importance of forensic engineering practice on the lives and safety of the general public, most of whom have only the highest expectation of the competency of the engineers responsible for the built environment and of the safety and durability of the built works.

It is for the benefit of the people who daily use the artifacts of civilization that civil engineers and forensic engineers involved in the investigation and evaluation of the built environment provide their services.

Chapter 1
Introduction

Failure can be defined as an unacceptable difference between an actual condition or performance and the intended or reasonably anticipated condition or performance. Failure need not involve a complete or even partial collapse; failure may involve a less catastrophic deficiency or performance problem, such as unacceptable deformation, cracking, water- or weather-resistance, or other such phenomena. The role of the forensic engineer may be that of an investigator of the causes of the failure, or it may extend into litigation support and testimony in legal proceedings. Forensic engineers are also involved in devising repairs or mitigating consequences of failures, however, the *Guidelines* do not intend to address issues of failure mitigation or repair.

Etymologically, the word "forensics" comes from the Latin *forensis* (of the forum)* and applies to the public forums of discussion, debate, or the administration of justice. College and high school debate teams are often referred to as "forensic societies." Although rooted in argument, the term has also become commonly associated with applying scientific or engineering knowledge or expertise to investigation. Here too, however, the term is generally associated with criminal investigation, or with some other domain related to our legal system. Formally then, forensic engineering can be defined as applying engineering principles, education, and knowledge to problems where legal liability is to be decided in a legal forum. However, much of the work of forensic engineers leads to the resolution of a problem without formal legal proceedings, and the term "forensic engineering" more generally and most often applies to studying problems that *may* be decided in a legal forum. The possibility that a forensic engineer will be retained as a consultant or expert in a legal proceeding distinguishes forensic engineering from other engineering disciplines, and the context of this naturally adversarial environment affects how forensic engineering is practiced.

For forensic engineers to intelligently investigate the cause of a failure, to help identify parties responsible for the failure, and to assist in mitigating the effects of the failure, they have to understand technical issues, and the professional, ethical, legal, contractual, and business practices that may have played a part in the failure. When forensic engineers possess that understanding, they can be of significant help in the effort to know what in fact failed, where, when, how, and why a failure has originated, and who may have contributed to responsibility for the failure. Because deficiencies and failures of the built environment can lead to claims, disputes, and legal proceedings, forensic engineers need to have some familiarity with relevant legal processes, and need to know how to work effectively with attorneys. One feature of the

* Retrieved from www.perseus.tufts.edu

Guidelines is its coverage of a range of subjects, the understanding of which is necessary for the effective and successful practice of forensic engineering.

Engineering outside of the forensic arena is, at its best, an exercise in addressing the concerns and accommodating the positions of many diverse parties involved in conceptualizing, planning, designing, constructing, operating, maintaining, renovating, reconfiguring, and decommissioning an engineered facility. The goal of the engineering effort can be seen as satisfying as many of those diverse parties in each of their areas of concern, or at least dissatisfying them as little as possible. The forensic arena, however, often involves an adversarial process concerned with a particular aspect of the facility or with a specific occurrence, where there can be "winners" and "losers." Because of that, some practicing design engineers may be of the opinion that a forensic engineer's role in helping identify responsible parties may itself border on unethical behavior, because it may merely be seen as a venal participation in the "blame and shame game." The Committee does not hold that opinion. In a well-ordered, just and caring society, identifying sources of failures, and holding responsible those who contribute to causing damage are worthwhile endeavors. Forensic engineers can aid in the processes of correctly characterizing the failure and of allocating responsibility.

The purpose of this publication is to document current practice and to provide guidance for the effective and ethical practice of forensic civil engineering. It is intended as a primer for the novice and as a reference for the seasoned practitioner.

No single author would have the details and the credibility to write on all the topics that were necessary for the *Guidelines*; therefore, many of the members of the Forensic Practices Committee were involved in the preparation of the individual chapters. The *Guidelines* chapters each deal with an important aspect of forensic engineering practice, such as competencies of a forensic engineer; the standard of care of engineers, including forensic engineers; the investigation and reporting processes; ethics; the legal forum; and business considerations.

Chapter 2: "Competencies and Qualifications of Forensic Engineers," describes the attributes and competencies that the Committee sees as appropriate for forensic engineers. The chapter brings up the problematic subjects of advocacy, objectivity, and bias, which are also addressed in subsequent chapters.

Chapter 3: "The Standard of Care," discusses that concept and explains that it is best understood as the boundary between negligent error and non-negligent error. The chapter shows that the standard of care is not a "Standard" in the typical engineering sense of the word, but that it is a duty of every engineer, including forensic engineers, to exercise care. The concept of care is discussed, as well as the necessary components of a valid and reliable method an expert can use to support testimony concerning the standard of care.

Chapter 4: "Investigations and Reports," describes typical steps in a forensic investigation and what the Forensic Practices Committee sees as constituting appropriate investigation procedures. The chapter addresses the collection and handling of evidence in legal procedures, describes proper use of Codes and Standards in investigation and analysis, and outlines the process of investigation, from initial client contact and conflict check through production of a report of the forensic engineer's opinion.

The sections in Chapter 4 dealing with Reports describe reasonable content and organization of written reports of forensic investigations and touch on some of the requirements of a report in the legal forum.

Chapter 5: "Ethics" is an important focus of the *Guidelines*. A primary objective of ASCE, TCFE, and the Forensic Practices Committee is the promotion of ethical behavior on the part of forensic engineers. The chapter cites the ASCE Code of Ethics as it applies to forensic engineers. The chapter also continues the discussion of advocacy, objectivity, and conflicts of interest, along with other ethics topics such as those related to expert opinions, fees, time constraints, reporting unethical practices, and other sanctioning processes.

Chapter 6: "The Legal Forum" gives a brief overview of the US court system as it relates to the design and construction industry. The chapter defines the role of the forensic engineer as an expert witness, and discusses the interactions of the forensic engineer with plaintiffs, defendants, attorneys for each side, and other witnesses. It discusses the role of the forensic engineer in non-adjudication forums and describes rules of evidence and case law that affect the qualification of expert witnesses and admissibility of expert testimony.

Chapter 7, "The Business of Forensic Engineering," describes the non-technical management and business sides of forensic engineering practice. The chapter discusses marketing of forensic engineering services within acceptable ethical limits and addresses the topics of contracts, insurance, and compensation for services.

The members of the Forensic Practices Committee believe that this Second Edition of the *Guidelines of Forensic Engineering Practice* will be used as much as, or more than, the first edition was, and that it will contribute to the improvement of the effective, ethical, and rewarding practice of forensic engineering.

Chapter 2
Competencies and Qualifications of Forensic Engineers

2.1. INTRODUCTION

As society and technology become more complex, engineered systems can fail in more complicated ways and due to combinations of apparently obscure reasons. Investigation, evaluation, and litigation involving highly complicated topics have led to the need for experts with specialized technical knowledge who can skillfully explain that knowledge and effectively offer relevant opinions. Forensic engineers and engineer expert witnesses play a significant role in investigating failures; they may present the findings of their investigations in court. In the litigation process, the trier of fact is asked to rely upon opinions of witnesses who have been admitted as experts. The competencies and qualifications of forensic engineers, therefore, are important topics for the *Guidelines* and are the subjects of this chapter.

Individual states of the US define what constitutes engineering and define who qualifies to practice engineering. States generally require engineers to practice only in their areas of competency; in addition, several states separately license particular engineering disciplines. This chapter will not examine or enumerate the different approaches to engineering licensure followed by the various states, or the qualifications for licensure of the states. This chapter summarizes the Committee's current understanding of what competencies a forensic engineer and an engineer expert witness could have, and of what abilities may be worthwhile for a forensic engineer and an engineer expert witness to possess. As with material in other chapters, attributes and competencies described in this chapter are intended as guidance rather than as a Standard for establishment or measurement of attributes or competencies of forensic engineers.

The issues of objectivity and advocacy often arise in discussions of the performance of forensic engineers as expert witnesses. This chapter articulates those discussions with regard to the role of the expert.

2.2. QUALIFICATIONS OF AN EXPERT

In the legal setting, courts have requirements for qualifying expert witnesses. More detailed information regarding the forensic engineer expert in the legal forum is presented in Chapter 6. It is not the duty or responsibility of the engineering profession to "qualify" practitioners as forensic engineers. It is, however, the duty of engineers to report unethical behavior, and that topic is addressed in Chapter 5 on Ethics. As will be discussed in more detail in Chapter 6, Federal Rule of Evidence 702 (2011) includes:

> *A witness who is qualified as an expert by knowledge, skill, experience, training, or education may testify in the form of an opinion or otherwise if:*
>
> 1. *the expert's scientific, technical, or other specialized knowledge will help the trier of fact to understand the evidence or to determine a fact in issue;*
>
> 2. *the testimony is based on sufficient facts or data;*
>
> 3. *the testimony is the product of reliable principles and methods; and*
>
> 4. *the expert has reliably applied the principles and methods to the facts of the case.*

2.3. ATTRIBUTES OF EXPERT ENGINEERS

Curricula and degree programs in expert engineering or forensic engineering are not generally offered in schools of higher learning, so forensic engineering competency is not achieved by obtaining a specific academic degree. Aside from admission by a court as an expert, being an expert requires competencies in the subject area in dispute as well as competence in forensic engineering and litigation support activities that allow the engineer to form and effectively present reasonable, reliable, and valid opinions. This chapter presents the Committee's understanding of competencies and attributes that are common among effective professional forensic engineers.

The following are possible indices of baseline competency in forensic engineering in a particular subject area:

- An undergraduate degree in engineering, preferably from a program recognized by the Accreditation Board of Engineering and Technology/Engineering Accreditation Commission (ABET/EAC)
- Technical ability in the subject area demonstrated by the engineer's own professional work, including design, research, teaching, or writings.
- A graduate degree in the subject area.
- Professional license or registration to practice in the subject area.
- Full-time practice in the subject area.
- Authorship of recent peer-reviewed publications on the subject.
- Authorship of textbooks on the subject.
- Active involvement with professional organizations that serve to advance the subject area.

- Awards or peer recognition for accomplishments in the subject area.
- Previous occasions of being accepted by other courts as an expert in the subject area for both sides of disputes.

The key attributes of an expert engineer are knowledge, skill, experience, education, and training. Forensic engineers with these attributes, who possess an appropriate subset of the characteristics itemized in the bulleted list above, and who can effectively perform their work accurately, objectively, and in a professional and unbiased manner would most likely capably serve as experts in their field.

2.4. COMPETENCIES OF FORENSIC ENGINEERS

Before an engineer is admitted as an expert by the court, the court will have determined that the trier of fact needs the input from the engineer because the trier of fact lacks the specialized knowledge which the engineer possesses and that the engineer's specialized knowledge is necessary for the trier of fact to make a decision in the case. The forensic engineer serving as an expert witness should have competence in explaining the technical, truthful evidence in such a manner that the trier of fact, the lay jury, or the judge can understand it and answer the ultimate question.

Forensic engineers not serving as expert witnesses also should have competence in explaining engineering opinions and conclusions. Forensic engineers, whether serving as expert witnesses or not, should have competence in investigating and analyzing the technical subject of their assignment and in reporting their observations, opinions, conclusions, and recommendations. Investigation and reporting are addressed in Chapter 4.

In addition to the above, a forensic engineer, including those serving as expert witnesses, should have competencies specific to the forensic and expert field. Some of those competencies are described in this section.

2.4.1. Understanding the Problem in the Context of the Investigation or Dispute

Engineers are often described as problem solvers, a role for which engineers are justifiably proud to be recognized. Forensic engineers often are asked to help solve a problem; however, a valuable competency of the forensic engineer is to understand not only what the problem is and what an effective solution might be, but also how a particular solution may affect parties to a dispute arising from the problem.

On the witness stand, the expert should only answer the questions that are asked, and not elaborate beyond that. An old joke demonstrates the importance of this concept:

> *Three people were going to the guillotine, and the first was the lawyer, who was led to the platform and blindfolded and put his head on the block. The executioner pulled the lanyard and nothing happened. So, out of mercy, the authorities allowed the attorney to go free. The next man to the guillotine was a physician, and he laid his head on the block, and they pulled the lanyard ... nothing. The blade didn't come down. So, to be fair, they let him go too. The third man to the guillotine was an engineer. He was led to the guillotine, and on the way he looked up at the release mechanism securing the blade at the top of the track. As he laid his head on the block, he said, "Wait. I think I see your problem."*

The point is that the effective forensic engineer and expert witness should understand the problem and its possible solutions in the context of the investigation or dispute. The engineer in the joke was not asked to solve the problem of the recalcitrant guillotine blade, and volunteering to solve it worked against his own self interest.

2.4.2. Understanding the Concept of Advocacy

Many references and articles state that the expert witness is not an "advocate" for the client. To the extent that "advocacy" entails deception, obfuscation, or denial of technically truthful evidence, this is certainly true. However, Fundamental Canon 4 of the ASCE Code of Ethics (2006) states, "Engineers shall act in professional matters for each employer or client as faithful agents or trustees, and shall avoid conflicts of interest." This canon acknowledges the fact that engineers are retained by clients for the purpose of providing a service to the benefit of that client.

Engineers also have an overriding obligation, typically under the terms of the licensing laws of the states in which they practice, and as emphasized in Fundamental Principle 1 of the ASCE Code of Ethics, to protect public safety and welfare.

These two tenets are not necessarily in conflict with each other, and a particular competency of an effective forensic engineer or expert engineer witness is to navigate an intricate path that accommodates the two.

It is the obligation of the expert engineer witness to provide truthful, adequately researched testimony in the best possible light for the client who retained the engineer. If, prior to testimony, the forensic engineer concludes, again through adequate and thorough development and evaluation of evidence, that the client's position is weak, wrong, or at substantial risk of being successfully challenged, the engineer is obliged (by the ethical duty to serve the client) to inform the client of that conclusion. If requested, the forensic engineer may advise and assist in settling the dispute, aiming to achieve the best possible outcome for the client despite the unfavorable circumstances.

2.4.3. Understanding Objectivity

Joseph S. Ward, P.E., stated (Ward 1989), "[T]he expert witness must be unbiased and objective in both the investigation and in testimony on the witness stand." Ward also wrote, "There are often many contributing factors to a failure and therefore conflicting opinions are plausible. Each forensic engineer may approach an investigation from a different viewpoint and two or more recognized experts may not always come to the same conclusion (Ward 1986)."

Kenneth Carper has written (Carper 1989), "Forensic engineers...must remain impartial seekers of the truth, even when the truth is not in their client's interest. The forensic engineer is required, by ethical principles, to maintain objectivity." Carper also wrote (Carper 1989), "Dissenting viewpoints among competent forensic engineers are common. Disagreement does not necessarily imply dishonesty or incompetence on the part of one of the witnesses. Often there is no single truth. Failures and accidents often result from a number of complex, interrelated factors. An honest expression of diversity of opinion, through the introduction of testimony by multiple experts, is a healthy approach to seeking truth. It is important to give consideration to all contributing factors."

These statements by acknowledged authorities in the field of forensic engineering illustrate the fact that objectivity does not necessarily result in experts on both sides of a dispute arriving at the same opinion. The expert for each side is retained by an attorney who is an advocate for the litigant, and each side's expert helps develop and support legitimate lines of argument that substantiate the claims or defenses of each party, to the extent that valid, reliable, truthful technical evidence exists. As consultants, competent forensic engineers should inform the attorney who retained them of all the strong and weak points of both sides to the extent that they are aware as a result of their investigation and analysis. However, as a disclosed expert providing testimony, the forensic engineer has a different role. The testifying expert's role is to assist the trier of fact to understand the evidence or to determine a fact in the issue. It is only through examination by an attorney, however, that an expert presents testimony. Even then, the testimony is only in response to the questions asked.

2.4.4. Understanding Bias

Bias is an unfair preference for a particular position, or the distortion of results due to a condition not appropriately taken into account. Bias is an inherent characteristic of human behavior. Understanding bias and overcoming its influence are important competencies of forensic engineers and expert engineer witnesses.

2.5. GENERALIST FORENSIC ENGINEER

Many failures require investigations be conducted by more than one specialist. Because clients or counsel may not be equipped to mobilize and coordinate investigations by several engineering disciplines, some cases involve the services of a principal investigator, or generalist forensic engineer. This person serves as the lead consultant who advises the client on the types of specialists needed for the team, selects them with the client's authorization, and coordinates their activities. The generalist may get involved in testimony if qualified to testify in any of the disciplines, or may simply observe the proceedings after coordinating the investigations and assisting with trial preparations.

To effectively serve as a lead consultant in this context, the generalist forensic engineer should have competence in organizing, managing, coordinating, overseeing, and administrating the specialists.

2.6. CONCLUSION

Education, training, experience, skill, and knowledge are attributes of effective forensic engineers and expert engineer witnesses. Advocacy, objectivity, and bias must be understood and appreciated in the conduct of an investigation and in legal proceedings. Truthful opinions, arrived at through thorough, valid, and reliable means of investigation and evaluation of technical data are an essential aid to the trier of fact.

2.7. REFERENCES

ASCE. American Society of Civil Engineers. (2006). Code of Ethics, ASCE, New York, Adopted July 23, 2006.

Carper, K. L., (Ed.). (1989). *Forensic Engineering*, Elsevier, New York.

U.S. Government Printing Office. (2011). *Federal Rules of Evidence*, Washington, DC.

Ward, J.S. (1986). What is a Forensic Engineer?, *Forensic Engineering: Learning From Failures*, American Society of Civil Engineers, New York.

Ward, J.S. (1989). The Engineer as Expert Witness, *Forensic Engineering*, Elsevier, New York.

Chapter 3
The Standard of Care

3.1. INTRODUCTION

Forensic experts are asked to consult and to provide expert testimony regarding technical issues in disputes involving failures and defects, such as the causes of a failure, the origin of a defect, the results of a failure or defect, and possible ways to affect repairs necessitated by a failure or defect. Forensic experts are also asked to provide testimony to assist the courts in assigning responsibility for damages arising from a failure or defect.

Among the most critical of the areas of testimony, certainly from the perspective of a defendant engineer, is the presentation of opinion concerning the "standard of care" and whether the defendant met that level of performance. The evaluation of an engineer's performance, the definition of the duty the defendant owed the plaintiff, and the comparison of the defendant's performance with that level of care are all questions of fact an expert witness might address. A forensic witness can be examining an engineer's performance as an expert for that engineer's attorney, or as an expert for the attorney of another party to the dispute. Regardless of for whom the expert is providing forensic engineering services, if the expert is to develop opinions concerning the performance of another engineer, the expert must understand the concept of the standard of care and use valid and reliable methods to come to and support opinions concerning the standard of care.

This chapter will discuss the standard of care and the necessary components of a valid and reliable method an expert can use to support testimony concerning the standard of care. As will be seen below, the standard of care is the measure of acceptable performance of a professional. The question of that performance can be raised in reference to an engineer's practice of any engineering discipline; it can also be raised in reference to a forensic engineer's performance as an expert or consultant. This chapter is intended to provide guidelines for the forensic expert investigating an engineer's role in a technical failure or defect. Additionally, because a forensic engineer can be held to the standard of care of forensic engineers, this chapter is also intended to assist in the forensic engineer's understanding of the standard of care and to appreciate that it is also a measure of the performance of forensic engineers.

A forensic engineer can be investigating or testifying regarding a practitioner's adherence to a standard of care and can also be held to the standard of care of forensic engineers that provide such investigation or testimony services. This chapter is intended to address both of those aspects of the standard of care.

Negligent error can lead to liability on the part of the perpetrator of the error. Error that has been shown to be non-negligent, that is, error a normally competent practitioner perpetrates despite the application of care, diligence, and judgment, should not lead to liability on the part of the practitioner. When a professional engineer is thought to be responsible for a failure and becomes a potential defendant in a dispute, forensic consultants and experts working for all sides in the dispute should understand what the standard of care was that the defendant should have met and understand what the defendant did relative to the standard of care.

3.2. THE STANDARD OF CARE

Engineers have a duty to provide their services in a manner consistent with the standard of care of their professions. A good working definition of the standard of care of a professional, derived from case law (*City of Eveleth* 1974; *City of Mounds View* 1978; *Gagne* 1934, and others), was presented in a standardized jury instruction, Book of Approved Jury Instructions, 6.37, "The Duty of a Professional," (BAJI 1986), which read:

> *In performing professional services for a client, a professional has the duty to have that degree of learning and skill ordinarily possessed by reputable professionals, practicing in the same or similar locality and under similar circumstances. It is his or her further duty to use the care and skill ordinarily used in like cases by reputable members of his or her profession practicing in the same or similar locality under similar circumstances, and to use reasonable diligence and his or her best judgment in the exercise of professional skill and in the application of learning, in an effort to accomplish the purpose for which he or she was employed. A failure to fulfill any such duty is negligence.*

The Book of Approved Jury Instructions was revised since that 1986 publication, and in fact is no longer typically used in California. In its place, The Judicial Council of California has published *Judicial Council of California Civil Jury Instructions,* (2011) which includes the following:

600. Standard of Care

> *[A/An] [insert type of professional] is negligent if [he/she] fails to use the skill and care that a reasonably careful [insert type of professional] would have used in similar circumstances. This level of skill, knowledge, and care is sometimes referred to as 'the standard of care.*
>
> *[You must determine the level of skill and care that a reasonably careful [insert type of professional] would use in similar*

circumstances based only on the testimony of the expert witnesses [,including [name of defendant],] who have testified in this case.]

It is important to appreciate that case law says a professional such as an engineer is not obliged to be perfect. Some deviation from perfection (i.e., some level of error) is to be expected of an engineer and accepted by the client and by society. An engineer can be deemed liable (obliged to pay) for damages arising from the engineer's professional negligent errors, which are those beyond or worse than an acceptable level of error. The engineer, however, is not liable for damages arising from that engineer's non-negligent errors, which are those errors within an acceptable range of deviation from perfection. The standard of care is the boundary between a non-negligent error and a negligent error. Although this boundary is a sharp line, that is, one is either on the negligent side or the non-negligent side, the placement of the line is not fixed. The boundary is determined in every instance when a jury answers the question, "Was this engineer negligent?"

3.3. EXPERT TESTIMONY

According to U. S. Federal Rule of Evidence 702—informed by case law such as *Daubert* (1993) and *Kumho* (1999)—the purpose of expert witness testimony is to "assist the trier of fact to understand the evidence or to determine a fact in issue." Expert testimony is admissible in Federal Court if 1) the testimony is based upon sufficient facts or data, 2) the testimony is the product of reliable principles and methods, and 3) the witness has applied the principles and methods reliably to the facts of the case (Federal 2001). Some state jurisdictions have not embraced the Daubert and Kuhmo standard for admissibility, but instead follow the standard defined in *Frye* (1923), which requires that the expert's testimony "must be deduced from a well-recognized scientific principle or discovery, and the thing from which the deduction is made must be sufficiently established to have gained general acceptance in the particular field in which it belongs." One application of the "general acceptance" test requires the methods used by an expert to support an opinion be those methods that are generally accepted within the relevant scientific community.

To assist the trier of fact (the jury or the judge) to determine the standard of care and the decision regarding the defendant's conformity to the standard of care, it is appropriate and necessary for an expert to testify concerning the engineer's duties. The jury should be aided in determining the facts by answering the following questions:

- Does the defendant have the degree of learning and skill ordinarily possessed by reputable professionals, practicing in the same or similar locality, and under similar circumstances?

- Did the defendant use the care and skill ordinarily used in like cases by reputable members of his or her profession practicing in the same or similar locality under, and under similar circumstances?
- Did the defendant use reasonable diligence?
- Did the defendant use his or her best judgment?
- Did the defendant do all that in an effort to accomplish the purpose for which he or she was employed?

To answer those questions, the following also have to be determined:

- Who are "reputable professionals"?
- What are "same or similar" localities and circumstances?
- What is "skill"?
- What is "care"?
- What are "like cases"?
- What learning and skill are "ordinarily possessed"?
- What care and skill is "ordinarily used"?
- What is "reasonable diligence"?
- What is "judgment"?
- What is the defendant's "best judgment?"
- What was the purpose for which the defendant was employed?

3.4. DEFINITIONS

Any valid and reliable method an expert uses to support an opinion regarding the standard of care as defined above should answer the questions listed above. An early task for the expert is to describe or acknowledge as pertinent the following terms (Webster's 2004):

- **Learning:** Knowledge or skill acquired by instruction or study; formal education.
- **Skill:** Knowledge of the means or methods of accomplishing a task; the ability to use one's knowledge effectively and readily in execution or performance; technical expertness; proficiency; a learned power of doing a thing competently; a developed or acquired aptitude or ability.
- **Ordinarily:** In the ordinary course of events; to the usual extent; moderately; without distinction.
- **Reputable:** Enjoying good repute; of excellent reputation; held in esteem; estimable (worthy of esteem or respect, deserving good opinion); respectable (conventionally correct in conduct).

To "have that degree of learning and skill ordinarily possessed by reputable professionals" means to have received instruction, study, or formal education, and to have knowledge of means or methods, or to have developed or acquired an aptitude or ability to accomplish the engineering task to the usual extent that engineers conventionally correct in conduct and deserving of good opinion obtain or have in the ordinary course of events. This means the engineer does not have a duty to attain the highest level of academic achievement, nor to pursue the highest academic degree, nor to have extraordinary knowledge, aptitude, or ability. One indicator of adherence to this duty is possession of an engineering license. States that license or register engineers typically demand a minimum level of learning in the form of several years of work under the direct supervision of a licensed engineer, or the possession of a relevant degree from an appropriate accredited institute of higher learning. States establish the existence of minimum levels of aptitude and ability by requiring that licensees or registrants pass a written examination (Kardon 2003).

To "use...skill ordinarily used in like cases by reputable professionals" means to bring to bear the knowledge of engineering means or methods, or to exercise the aptitude or ability to accomplish the engineering task, to the usual extent that engineers conventionally correct in conduct and deserving of good opinion exercise in the ordinary course of events. This means the engineer is not bound to carry out every engineering task perfectly, or with an extraordinary level of skill, but has a duty to do so as is usually done, including an acceptable degree of mistake or error (Kardon 2003).

- Similar: Marked by correspondence or resemblance.
- Locality: Particular spot, situation, or location.
- Circumstances: Surrounding conditions.

The learning, care, and skill that are required are those possessed and used by reputable engineers "practicing in the same or similar locality and under similar circumstances." To evaluate whether an engineer met that duty, one must know relevant aspects of the locality of the engineer's practice and relevant circumstances under which the engineer's services were provided. Aspects of the locality and circumstances which may be relevant to the question of an engineer's actions include, but are not limited to, the local conventional materials and practices, the predominance of a particular soil type in the area, the state of the real estate market, the form of contract used, the architectural complexity of the structure, and the intended occupancy and use of the structure.

- **Care:** Joan Tronto (1993) identifies five elements of the ethic of care, all of which are pertinent in engineering (Kardon 2005). Tronto's five elements of care are:
    - **Attentiveness:** Caring about; noticing the need to care, or recognizing the need of others.
    - **Responsibility:** Taking care; having more than an obligation, but a responsibility arising in part from one's position or knowledge.

  - **Competency:** Care giving; having the ability to carry out the caring act effectively and correctly.
  - **Responsiveness:** Care receiving; being receptive to care, being aware of the care receiver's perceptions.
  - **Integrity:** Knowledge on the part of the caregiver of the situations, needs, and competencies of all the parties to the care interaction, including an assessment of needs in a social and political, as well as personal, context.
- **Reasonable:** Being or remaining within the bounds of reason, not extreme, not excessive, well balanced; sensible (having or containing sense or reason, characterized by or resulting from sober, serious examination and study).
- **Diligence:** Persevering application; devoted and painstaking application to accomplish an undertaking; assiduity (constant, unremitting attention or close, persistent, energetic application).

A professional has the duty to use "reasonable diligence" in the exercise of skill and learning. This means the engineer must apply a well balanced, sensible level of effort to the engineering tasks; the effort must involve or result from serious examination, but the effort need not be extreme, excessive, or beyond the bounds of reason. In evaluating the reasonableness of an engineer's diligence, one can examine the depth of the engineer's examination of loads, components, systems, and materials; the breadth of alternate design choices examined; the type and rigor of the investigation or of the structural design analysis methods used, and the suitability of the analysis method in consideration of the requirements of the structural system. It is difficult to define a reasonable level of diligence. Every time an engineer selects a design approach, chooses an analysis method, or selects a particular element or member of a structure or system to investigate or analyze, the engineer strikes a balance between doing too much and doing too little (Kardon 2003). Evaluation of diligence must take into account the concept that "rigor is not a substitute for imagination (Klein 1999)."

- **Judgment:** The mental or intellectual process of forming an opinion or evaluation by discerning and comparing; the power or ability to decide on the basis of evidence; the wise or just exercise of the capacity to judge; discernment; discretion.

Gary Klein describes the problem-solving abilities people bring to bear in making decisions as "sources of power" (Klein 1999). These abilities can be considered aspects of judgment. They include:

- Intuition (pattern recognition, having the big picture, achieving situation awareness).
- Mental simulation.
- Using leverage points to solve ill-defined problems.

- Seeing the invisible (perceptual discriminations and expectancies).
- Storytelling.
- Analogical and metaphorical reasoning.
- Reading peoples' minds (communicating intent).
- Rational analysis.
- Team mind (drawing on the experience base of the team).

Klein also found other judgments and abilities that apply to decision making:

- Judging the typicality of a situation.
- Judging typical goals.
- Recognizing typical courses of action.
- Judging the solvability of a problem.
- Detecting anomalies.
- Judging the urgency of a problem.
- Detecting opportunities.
- Making fine discriminations.
- Detecting gaps in a plan of action
- Detecting barriers that are responsible for gaps in a plan of action.

- **Best judgment:** A professional has a duty to "use his or her best judgment" in the exercise of skill and learning. This is the only obligation the professional has that is described in BAJI in the superlative. In evaluating the judgment of an engineer, one must refer to that particular individual's "best," not the best judgment of all engineers. To determine whether the individual's best judgment was used, one must examine the judgment actually brought to bear. An appropriate test of whether an engineer applied that engineer's own "best judgment" is to examine what that engineer knew "in his heart of hearts," in the seat of his truest feelings, was the right course of action.

Judgment is informed by knowledge gained from experience, and the best judgment of a particular engineer arises from that individual's own experience. To assess an engineer's best judgment, one must consider the knowledge and experience of that engineer, including knowledge gained from formal education, and examine how that engineer's judgment was applied in the situation in question.

- **Purpose:** An investigation of the purpose for which the defendant engineer was employed must take into account the specifics of the contract under which the defendant engineer provided services, and the scope of those services, including the modifications of that contract arising from the specific performance of the parties as well as from written or oral contract changes.

The investigation must also take into account the fact that engineers are hired to be engineers, licensed or registered by the states in which they practice for the purpose of protecting the public, and are bound by that license to act in a manner to fulfill that purpose. Additionally, the actions of engineers who are members of professional societies are defined or guided by those societies' codes of ethics, and members are asked to perform in accordance with those codes for the purposes stated in them (Davis 1998).

3.5. METHODS

Testimony concerning the standard of care should address each of the issues itemized above. In any one particular case, one or several of the issues might be moot or irrelevant or might not be evidence for which the court requires expert testimony. For instance, an expert witness may not be required to testify as to the level of learning a defendant possesses.

On the other hand, many of the above issues do reasonably require expert testimony. One such area of testimony is the comparison of the defendant's actions or performance with those of reputable professionals. Often, such testimony is based on surveys of other practitioners. The validity and reliability of a survey upon which testimony is based can be challenged if the survey population is incorrectly defined, if the survey sample is not appropriately selected, if the sample size is not statistically significant, or if the survey results are not statistically evaluated.

3.6. VALIDITY

There are two general types of validity: external and internal. External validity (Campbell and Stanley 1963) is the extent to which the method is generalizable or transferable. A method's generalizability is the degree the results of its application to a sample population can be attributed to the larger population. A method's transferability is the degree the method's results in one arena can be applied in another similar arena.

In contrast to external validity, internal validity "is the basic minimum without which the method is uninterpretable" (Campbell and Stanley 1963). Internal validity of a method addresses the rigor with which the method is conducted (e.g., the method's design, the care taken to conduct measurements, and decisions concerning what was and wasn't measured). There are different types of internal validity: face validity, content validity, criterion-related validity, and construct validity.

Face validity is the degree to which a method appears to be appropriate for measuring what it intends to measure (Fink 1995). An example of face validity is the observation that a ruler appears to be an appropriate tool to measure length.

Content validity has to do with the degree to which the method measures the trait it is intended to measure. An example of a test which lacks adequate content validity is one which intends to measure a subject's mathematical ability by testing only addition (Carmines and Zeller 1979).

Criterion-related validity has to do with the degree to which the method allows for assessment of a subject's performance in situations beyond the testing situation—in a different domain than the test. Criterion-related validity may be concurrent or predictive. That is, the test result may either be intended to assess a criterion independently measured at the same time (concurrent), or to predict achieving a criterion in the future (predictive). An example of predictive criterion-related validity is the extent to which a written driver's test accurately predicts how well the tested population will drive (Carmines and Zeller 1979). The written driver's test does not involve physically driving a car on the road, but only involves answering several multiple-choice questions. The extent to which good performance on that written driver's test correlates well with future good driving performance on the road is a measure of the test's criterion-related validity. The physical act of driving a car takes place in a different situation than the testing environment, and involves different skills and abilities. Threats to the predictive validity of the written driver's test include the possibility that a test subject can't read English but might be a good driver. That threat to criterion-related validity is addressed by having the test printed in several languages besides English. An inquiry of a test's predictive criterion-related validity asks the question, "How accurately does this test measure future performance in that setting?"

Construct validity has to do with the degree to which the results of the method can be accounted for by the explanatory constructs of a sound theory. A method's construct validity is understood by first specifying the theoretical relationships among the concepts, examining the empirical relationships among the measures of the concepts, and then interpreting how the observed evidence clarifies the concepts being measured (Carmines and Zeller 1979). Construct validity is demonstrated when measures that are theoretically predicted to be highly interrelated are shown in practice to be highly interrelated. An inquiry of a test's construct validity focuses on the question of whether the results of the test are in fact a true measure of the construct or theory being tested, and not of some other phenomenon or process which might produce the same results. Such an inquiry asks, "Is this theory the best explanation for the results?"

Not all types of internal validity are applicable to any one method (Rossi and Freeman 1979). For instance, a method may not be one which is intended to predict an outcome of a process; it may not seek to answer the question, "How accurately does the test measure future performance in a different setting?" so its criterion-related validity would not be an issue. The intended result of a method may not be something that is predicted by a theory; it may not ask, "Is the theory the best explanation for the results?" If the result of the method can't be measured and contrasted with a theoretically predicted result, its construct validity would not be an issue.

3.7. RELIABILITY

A reliable method is one that yields consistent results upon repeated use; it is suitable for its intended purpose. However, when a reliable method is used in court by experts on two opposing sides in a dispute, the results will not necessarily be identical. After all, the reason disputes end up in court is just that there are good arguments for both sides. Expert witnesses are ethically obliged to help their attorney clients explain the case to the juries from the particular point of view of their client, within the bounds of truth (Kardon, Schroeder, and Ferrari 2003). It is not unethical for an expert witness to explain technical aspects of the dispute from the particular point of view of the client. Issues end up in court because there are differences of opinions and shadows of doubt, and experts retained by attorneys representing both sides of a dispute often come to different and contrary opinions based on reasonable interpretations of the evidence they each review or develop. Each side's expert presents technical evidence for the purpose of aiding the trier of fact. The trier of fact is best served by the effective presentation of technical arguments from both sides of the dispute. The reliability of a method used as the basis of an opinion given as expert testimony, therefore, cannot be evidenced by identical opinions being supported upon its repeated use in the dispute resolution process. Instead, the reliability of such a method will originate in the understanding that the method is suitable for supporting the opinion of the expert.

3.7.1. Example: Invalid and Unreliable Survey of the Standard of Care

A real dispute involving alleged defective engineering design and construction of a single family house in Oakland, California, is an example of the way some structural engineer expert witnesses currently arrive at their opinions regarding the standard of care; it illustrates the problems with the current method. The case involved a house constructed on a vacant down-slope lot (Kardon 2001). The owner-developer of the house retained a structural engineer to provide engineering design services for the pier and grade beam foundation. Soon after construction was complete, the owner noticed cracks opening up in the foundation grade beams and in the interior walls and floors of the house. Measurements revealed the floors were as much as four inches out of level and the walls were as much as three inches out of plumb. The owner sued the structural engineer for negligence in his design of the foundation.

Investigation and testimony that took place in preparation for the trial revealed the following evidence:

- The structural engineer of record (SER) for the original design of the foundation had been in private practice for many years and had designed several pier and grade beam foundations.
- Most, if not all, of the foundations the SER had previously designed were for buildings on flat lots, with expansive or compressible soils; none was for a building on a sloping lot, where the effect of soil creep (the natural tendency of soil on a slope to gradually move downhill) and hillside stability was significant.
- The SER did not refer to any geotechnical engineering reports in his design of the foundation, did not consult with a geotechnical engineer during his design process, and did not recommend the owner retain a geotechnical engineer for the project.
- The SER's engineering calculations did not include any consideration of soil creep as a design load on the foundation, or consider ground stability. They indicated the SER's design intent for the foundation was to transfer the vertical dead and live loads from the building to the soil through skin friction or end bearing. Using an assumed value for skin friction, and an assumed depth "to bedrock" of 5 feet, the defendant SER's design called for the pier depth to be 10 feet.
- Soil explorations carried out in preparation for the trial identified resistant soil, appropriate for supporting foundations resisting creep forces and for providing stability on the hillside, at a depth of 10 feet.

The reason the house exhibited cracks and tilted floors and walls was that the foundation piers were not deep enough into resistant soil to develop fixity against creep of the surface soils and down slope movement, were not strong or stiff enough to resist creep forces without excessive cracking and deflection, and were not adequately connected to the grade beams to transfer moment and shear forces between the piers and the grade beams.

The structural engineer expert for the plaintiff owner was of the opinion that the SER was negligent in designing the foundation without accommodating the structural effect of soil creep. The expert was also of the opinion the engineer was negligent for carrying out the design of the foundation without the participation of a geotechnical engineer, considering the defendant's lack of experience with hillside foundations.

The structural engineer expert for the defendant SER, however, testified the SER "met the standard of care" and therefore was not negligent. The defense expert described his method to determine the standard of care as an informal telephone survey of eight structural engineers in the area who were practicing at the time the project was designed. The expert asked the survey respondents whether, in their practice of designing house foundations at the time the subject foundation was designed, they relied on the participation of a geotechnical engineer. Four respondents answered "yes" and four "no." Based on those responses, the expert came to his opinion that the SER defendant was not negligent in designing the foundation because his survey told him that, at the time of the project in question, structural engineers were as likely as not to have designed foundations without the participation of a geotechnical engineer.

The SER's expert did not ask the survey respondents whether any of the foundations they designed were on sloping sites or whether they considered the effect of soil creep on such foundations. He did not ask the respondents to characterize their level of experience with foundation design for houses on sloping sites. In the absence of these clarifications, answers to the survey questions did not reveal practices of structural engineers directly relevant to the defendant's actions, and therefore could not substantiate the standard of care. Because of these deficiencies, the method the defense expert used to determine the standard of care was invalid. Additionally, the defendant's expert did not define or select his sample population in any statistically meaningful way and carried out no analysis to predict how accurately the sample population's answers would match those of the overall population of structural engineers contemporaneously practicing in similar circumstances. Because of these failings, the method the expert used was unreliable.

3.7.2. Example: Expert Testimony Based on Invalid and Unreliable Methods

An expert testified on behalf of an insurance company concerning the amount of structural movement that must have occurred in a house that was in the throes of a major remodel and seismic upgrade when it was allegedly damaged by the 1989 Loma Prieta earthquake (Kardon 2002). The expert did not observe the actual damage caused by the earthquake, but was asked by his client to determine whether the damage that was claimed by the homeowner could have been caused by the earthquake. The expert performed a computer-based analysis of the house and relied on that analysis to come to his opinion regarding the amount of movement the house underwent during the earthquake and, therefore, the amount of damage to the house that occurred as a result of the earthquake.

The expert testified that his computer model was based on the assumption that the plywood for all the shear walls was in place at the time of the earthquake but was not nailed with all the nails specified in the design documents for the strengthening of the house. He testified that he assumed the plywood was sufficiently nailed so that there was continuity between the plywood and the framing. He assumed the plywood was sufficiently attached to the framing at the time of the earthquake because that was the assumption of the computer model. He also testified his model assumed none of the hold-down hardware was in place. He did not testify as to whether his model assumed the bottom plate nailing or the top plate shear transfer clips were in place.

The SER for the remodel and seismic upgrade, who performed structural observation during construction both before and after the earthquake, testified that most of the plywood on the exterior walls of the house at the time of the earthquake was attached with duplex nails (double-headed nails for easy removal) at a much wider spacing than the final shear wall nailing was to be, and the shear transfer clips and hold downs were not in place. The purpose of the plywood in place at the time of the earthquake was not to provide lateral load resistance, but to provide some jobsite security by preventing unauthorized access to the building after working hours.

There was no assurance that the analysis by the insurance company's expert accurately replicated the actual behavior of the building in the earthquake. This was because in finite element analysis, the real assembly of framing, plywood, nails, clips, hold downs, and other features that make up a real shear wall is modeled using elements of assumed strength, stiffness, and boundary conditions. These assumptions must be verified either by comparing the actual assembly to previously tested assemblies that have been shown to be accurately modeled, or by carrying out physical tests of the assemblies to compare their behavior with the model element's behavior.

Published models of wood-framed shear wall assemblies used in finite element analyses are based on assumptions of fully nailed walls, with competent, active shear and overturning transfer hardware in place. There was no testimony given by the insurance company's expert that he used a verified model of a wood-framed shear wall without hold downs and shear transfer hardware, and with plywood sheathing only lightly nailed at the panel edges.

In addition to the modeled wall assemblies, the interface of the house foundation and the ground, and the actual condition of the real foundation (cracks and all) must be accurately modeled in order for the analysis to be correct. There was no assurance that the soil–structure interaction and the foundation were accurately modeled.

The same criticism can be brought against the modeled loads. The insurance company's expert testified that he estimated the earthquake ground motion at the subject house by examining published earthquake ground motion records from two nearby seismometer stations, and used that assumed ground motion as input for his computer analysis of his model of the house. The actual ground motions to which the building was subjected must be accurately modeled in order for the finite element analysis to produce results which reliably duplicate the actual behavior of the building. The insurance company expert relied on records of earthquake motion recorded at two stations that he stated were close to the building site and that he stated were on similar soils. The insurance company's expert offered no testimony of any characterization of the soils at the subject site or of the sites where the earthquake motions he used were recorded.

Interpolation of earthquake ground motion at a particular site from ground motions recorded at other sites is inaccurate. The United States

Geological Survey (USGS 2012), the source of one of the ground motion records used by the insurance company's expert, publishes maps of earthquake-induced ground shaking. They state, "ground motions and intensities typically can vary significantly over small distances, these maps are only approximate. At *small scales, they should be considered unreliable.*" (Emphasis added.) By the same token, any ground shaking at the subject house that the insurance company's expert deduced from the two recording stations to which he referred must be viewed as unreliable. The only true measure of actual ground shaking at the site, absent a calibrated and functioning recording device, is the amount of damage that actually occurs at the site.

Because the model used by the insurance company's expert was not based on actual conditions at the building, and because no verification or justification of the model structure or loading was presented, the analysis was not valid or reliable. The computer model did not recreate an accurate depiction of the actual condition of the house or of the actual loads applied to the house by the Loma Prieta earthquake; it therefore lacked face validity. Because the model of the structure and of the loads was not representative of the real structure or loads, the method could not predict or describe the movement of the building, and therefore lacked criterion-related validity. Because of the absence of validity and reliability, it was argued that the expert evidence should be disallowed in the determination of damages.

3.8. A VALID AND RELIABLE METHOD

A method has been proposed (Kardon 2003) that does not rely on statistically questionable surveys, but is based on the deductive logic of case-based reasoning, which is an analogical reasoning technique that utilizes past problem solutions to solve current problems (Maher, Balachandran, and Zhang 1995). This reasoning technique has been used for two general tasks: diagnosing or interpreting a new situation in order to classify it as being included in or excluded from a group of known previous situations, and solving a problem by applying a previously discovered solution to a new situation (Leake 1996). The proposed method lends itself well to the first type, interpretive case-based reasoning. The basics of interpretive case-based reasoning involve the following steps (Leake 1996):

- **Assessment:** Characterizing the new problem situation.
- **Retrieval:** Identifying relevant past situations similar to the new one.
- **Comparison:** Using those past situations to suggest a classification of the new one.
- **Learning:** Updating the knowledge base with the learning from the new situation.

3.8.1. Assessment

The first step in the case-based reasoning method is "situation assessment" (Kolodner 1993), or the characterizing of the new problem situation. When an expert is asked to come to an opinion regarding an engineer's performance, the relevant and important details of that performance must be identified. In the above case study involving the failed foundation of a hillside house, the critical aspects of the defendant engineer's performance were his inexperience with designing such foundations and his lack of reliance on independent geotechnical engineering input given that inexperience. Also important were the descriptions of observed damage and an explanation of the causal relationship between the engineer's actions or omissions and the damage to the house.

3.8.2. Retrieval

Once the new problem situation is assessed and the important and relevant details of the situation and of the engineer's performance are characterized, the next step in case-based reasoning is retrieval, the identification and selection of relevant past situations similar to the new one. Sources of past situations can include the published record or situations known by the witness. In the proposed method, hypothetical cases formulated to illuminate a specific action or omission can also be included if there are no past situations to draw on, but only if the hypothetical is a fair representation of the performance of reputable engineers.

Relevance of a retrieved case can be justified by comparing its details to those of the subject situation. In the above case study, relevant retrieved situations would include cases in which engineers designed residential foundations and practiced contemporaneously with the defendant and in the same area where the defendant practiced. The retrieved situations should include cases in which engineers had varying levels of experience with hillside residential foundations and in which engineers provided services either with or without the participation of a geotechnical engineer.

In the above example, the defense expert could have carried out the retrieval task by asking the eight survey respondents questions that were relevant to the defendant's actions. Because case-based reasoning is not a statistical survey method, the proposed method does not involve a statistical selection of the survey population from the target population and does not rely upon a statistical verification that the survey answers replicate the answers of the target population. Case-based reasoning requires that the subject situation be correctly characterized, and the past situations must be relevant and comparable.

3.8.3. What is a "Case"?

Rosenman, Gero, and Oxman (1992) describe a "case" in case-based reasoning in design as containing three essential ingredients of an "entire design episode"—the problem, the solution, and the process for arriving at the solution. Every referenced design case includes the "problem," an expression of the design goals and requirements and the constraints on those requirements; the "solution," the results of all the design choices, the outcome of the design process, or a description of the physical manifestation of the thing designed, including the relationship of the thing to the other parts of the designed environment; and the "process," a description of the procedure of making and justifying design choices. Case-based reasoning in design is a form of the second type of case-based reasoning mentioned above: solving a problem by applying a previously discovered solution to a new situation. The method proposed here to serve as a basis for expert opinion, however, uses interpretive case-based reasoning to diagnose or interpret a new situation in order to classify it as being included or excluded from a group of known previous situations. The situations need not include a design solution. In fact, the situation examined is as likely as not to have nothing to do with a design.

This application of case-based reasoning will look at a "case" slightly differently. The purpose of the reasoning is to serve as a basis for an expert's opinion, which is the content of the expert's testimony. According to Rule 703 of the Federal Rules of Evidence, an expert witness may rely on facts not in evidence, including opinions of other experts with whom the testifying witness consults prior to the testimony. If those facts or data are "of a type reasonably relied upon by experts in the particular field in forming opinions or inferences upon the subject," the expert's testimony can be admitted. In relying on interpretive case-based reasoning as the basis for an opinion, the expert examines retrieved cases that may not have solutions in the form of adjudicated rulings of professional negligence. The solution or outcome of a retrieved case in this application is not, therefore, a determination of negligence or non-negligence in some other previous lawsuit. It is an opinion held by the witness or by those with whom the witness consulted. The "problem" part of a case used in this application is a set of circumstances containing an act or omission of an engineer. The "process" part of a case used in this application is the expert's or the consultants' reasoning or justification for the opinion.

3.8.4. Comparison

The comparison process in interpretive case-based reasoning involves the selection of past situations among those retrieved, which are proposed to be descriptive of the problem situation. The proposed situations are then justified as being descriptive by comparing their specific details with those of the problem situation. Differences between the proposed situations and the problem situation are examined to determine if those differences are significant enough to exclude the proposed situation from consideration as descriptive. Justification of the proposed descriptive situation can then be tested by critically examining its relevance and similarity to the problem situation and to similar hypothetical situations. The critical examination may indicate a necessity to retrieve different cases, in which case, the above steps should be repeated (Kolodner and Leake 1996).

It is possible that not all the characteristics of the problem situation are exactly matched by every retrieved situation (Rosenman et al. 1992; Kolodner 1993). In that case, and because "no old situation is ever exactly the same as a new one" (Kolodner and Leake 1996), retrieved situations can be adapted by substituting or transforming a feature to result in a "better fit for the problem description" (Kolodner and Leake 1996). How different from the problem situation a retrieved case is before it is considered irrelevant, and how appropriate the adaptation of a retrieved case is, are assessments the investigator must make and document. They are also matters to be presented to the jury for consideration.

In the above case study, a comparison of the past situations with the problem situation should have included an examination of whether engineers with varying levels of experience designing hillside residential foundations differed in their practice of working with geotechnical engineers. The way creep forces were handled in the past situations should be examined as well.

3.8.5. Learning

The fourth step of interpretive case-based reasoning is updating the knowledge base with the learning from the new situation. Creation and maintenance of a formal, active, growing database of engineering cases for the purpose of informing investigations into the standard of care is a desirable goal. There have been databases of engineering cases, such as the Architecture and Engineering Performance Information Center (AEPIC) (Loss 1987), which was started in 1982 at the University of Maryland; the Center for Case Studies in Engineering, which had its start in 1964 as the Engineering Case Program at Stanford University with support from the National Science Foundation, and continued at Rose-Hulman Institute of Technology with support from the American Society for Engineering Education (ASEE) (Engineering 1993); and the collection at the Online Ethics Center for Engineering and Science (Online 1995). These collections of engineering cases (and others) have been compiled for the purpose of aiding and studying the engineering design process or of studying engineering ethics. They were not established to document engineering practices in order to serve as a basis for opinions concerning the standard of care; therefore, they may not have sufficient detail to inform those opinions.

3.9. THE DELPHI PROCESS OF EXPERT CONSENSUS

In many instances, case-based reasoning will be sufficient for an individual expert witness to use in independently arriving at opinions regarding the performance of a defendant engineer. In some situations, however, such as those in which hypothetical situations are relied upon, where the problem situation is complex or unique or where the question of a defendant's negligence hinges on arcane or highly technical facts, the validity and reliability of the case-based reasoning approach may require more than the knowledge and experience of the individual expert witness. In those instances, the proposed method utilizes the Delphi process to support case-based reasoning.

The Delphi process, developed in the early 1950's by the RAND Corporation (Dalkey and Helmer 1962), was devised as a means to obtain a consensus of expert opinion concerning predictions of complex, uncertain future events. It is "a means for aggregating the judgments of a number of individuals in order to improve the quality of decision making" (Delbecq, Van de Ven, and Gustafson 1975). In the proposed method, it will be used to obtain a consensus of opinion concerning the characterization of the problem situation, the identification of relevant past situations, and (or) the suitability of the comparison between the problem situation and the relevant past situations.

The Delphi process involves a panel of various qualified respondents fairly selected by an investigator. The process is based on the panelists individually answering a series of questionnaires. The Delphi panelists do not meet face-to-face and remain anonymous to each other to foster independent thought, to eliminate peer pressure, to eliminate the possibility of one respondent deferring to another because of reputation, to eliminate a respondent answering in deliberate contradiction to another because of personality differences, or to reduce a respondent's devotion to an opinion simply because it was publicly expressed (Brown 1968; Turoff and Hiltz 1996). The first questionnaire is formulated to elicit answers representing a wide range of opinions concerning the subject of the investigation.

In the second questionnaire, all the answers of the first questionnaire are distributed (anonymously) to all respondents. Respondents are asked to reconsider their answers to the first questionnaire in light of all the other respondents' answers and to express the reasons for their answers. The third questionnaire distributes the answers and the descriptions of the reasoning the respondents provided in the second questionnaire, and elicits from the respondents criticism of the others' answers and reasoning. The fourth questionnaire distributes the collected criticisms and asks the respondents to consider those criticisms and express their final answers. Respondents are encouraged to modify their earlier opinions in light of the collected answers and criticisms, if those responses were persuasive.

The investigator collects and analyzes answers to the fourth questionnaire and identifies a consensus opinion. The investigator can then refer to this consensus in applying the case-based reasoning approach to determining the standard of care and in assessing a defendant's level of care.

The Delphi process has been criticized for its questionable adherence to the "basic, minimum standards applicable to all empirical science" (Sackman 1975). The reliability and validity of the Delphi process is questioned due to an absence of adequate verification and control. Sackman states, "[The Delphi Method, like any questionnaire,] is reliable and valid only to the extent that it is administered under conditions that replicate the basic experimental controls under which it was originally designed, tested, and validated" (pg. 12). In the proposed use of the Delphi process, the consensus opinion derived from Delphi is not intended to substitute for the judgment of the expert; the expert uses the consensus as substantiation of one or more of the steps in case-based reasoning: assessment, retrieval, and comparison.

The standard of care is not necessarily a statistical measure of engineering performance. It is instead an assessment of the quality of care exercised by a practitioner, with specific importance assigned to the engineer's application of judgment, diligence, and the five elements of care. Validity of the Delphi process in substantiating the case-based evaluation of an engineer's achievement of an acceptable level of care arises from the relevance of the questions, the qualifications of the panelists, and the reasonableness of the investigator's interpretation of the Delphi results.

Selected panelists should have a range of experience and expertise that encompasses that of the engineer whose performance is being investigated. The panel should not be made up entirely of experts with knowledge and experience beyond the subject's (Ziglio 1996), because the subject is to be compared to "reputable" or "reasonably careful" professionals not "experts," and it should be they who inform the investigator's opinions regarding the care, best judgment, and reasonable diligence of the subject engineer.

The Delphi process can be applied at any step of the case-based reasoning that is used as a basis for the expert's opinion. The Delphi panelists can be asked to characterize or assess the problem incident; they can be asked to define or select cases for retrieval, or to justify or adapt retrieved cases; or they can be asked to compare the problem incident with the retrieved cases. A Delphi process would be used whenever the expert believes the case-based reasoning needs that external substantiation, perhaps because of the nature and details of the problem situation, because of knowledge of the opposing side's strengths or weaknesses, or because of the scope of the expert's own experience.

3.10. CONCLUSION

The standard of care is an essential concept in engineering practice. As the boundary between negligence and non-negligence, it is a measure of an engineer's performance whenever that performance is called into question in a court of law. In court, testimony concerning the standard of care and a defendant engineer's performance relative to that level of care is presented by expert witnesses qualified by the court to give such testimony. In addition to qualification of the expert, the testimony itself is subject to a test of admissibility. Judges may disallow expert testimony, including that which provides evidence concerning the standard of care, if they determine there is not a valid or reliable basis for the evidence.

One critical measure of the standard of care lies in the evaluation of the "care" which the engineer brought to bear in a given situation. The ethic of care as defined by Tronto comprises elements that describe and define care. Each of the five elements can be assessed at every phase of the engineering process, including every phase of the forensic engineering process.

Engineers' errors and mistakes can cause injury to others. However, some of those errors and mistakes are to be expected, due to the nature of the engineering activity and of the systems within which the engineering activity takes place. The law allows the engineer some relief from liability for damage found to be caused by the engineer's non-negligent errors. The standard of care, the boundary between negligence and non-negligence, is a measure of an engineer's performance, and has a significant impact on the lives and livelihoods of everyone interacting with the built environment.

Engineers must exercise care in carrying out their services. Engineers have a duty to use the skill and care that a reasonably careful engineer does, which can include the duty to be reasonably diligent, to bring to bear their own best judgment, and to apply skill and care (attentiveness, responsibility, competency, responsiveness, and integrity) in an effort to accomplish the purpose for which they are employed.

3.11. REFERENCES

BAJI (Book of Approved Jury Instruction). (1986, January). 6.37, Duty of a Professional, *California Jury Instructions, Civil,* West Publishing Company.

Brown, B.B. (1968). *Delphi Process: A Methodology Used for the Elicitation of Opinions of Experts,* Paper P-3925, The RAND Corporation, Santa Monica, California.

Campbell, D.T., Stanley, J.C. (1963). *Experimental and Quasi-Experimental Design for Research,* Houghton Mifflin Company, Boston.

Carmines, E.G., Zeller, R.A. (1979). *Reliability and Validity Assessment*, Sage University Paper Series on Quantitative Applications in the Social Sciences, John L. Sullivan, Ed., Series No. 07-017, Newbury Park, California.

City of Eveleth v. Ruble, 302 Minn. 249, 253, 225 N.W.2d 521, 524 (1974).

City of Mounds View v. Walijarvi, 263 N. W. 2d 420, 424 Minn. (1978).

Dalkey, N., Helmer, O. (1962). *An Experimental Application of the Delphi Method to the Use of Experts,* Memorandum RM-727-PR (Abridged), The RAND Corporation, Santa Monica, California.

Daubert v. Merrill Dow Pharmaceuticals, Inc., 509 U. S. 579, 113 S. Ct. 2786, 125 L. Ed. 2d 469 (1993).

Davis, M. (1998). Thinking Like an Engineer: Studies in the Ethics of a Profession, Oxford University Press, Oxford.

Delbecq, A.L., Van de Ven, A.H., Gustafson, D.H. (1975). *Group Techniques for Program Planning: A Guide to Nominal Group and Delphi Processes*, Scott, Foresman and Company, Glenview, Illinois.

Engineering Case Library Internet Catalogue. (1993). The Center for Case Studies in Engineering, Rose-Hulman Institute of Technology, Terre Haut, Indiana ftp://alfred.carleton.ca/pub/civeng/ECL/cat-f93.txt.

Federal Rules of Evidence, U.S. Government Printing Office, Washington, DC, (2001).

Fink, A., (Ed.). (1995). *The Complete Survey Research Kit*, Volumes 1-9, Sage, Thousand Oaks, California.

Frye v. United States, 54 App. D. C. 46, 293 F.1013 (1923).

Gagne v. Bertran, 43 C. 2d 481, 275 P. 2d 15 (1934).

Judicial Council of California Civil Jury Instructions (CACI), LexisNexis Matthew Bender, (2011).

Kardon, J.B. (2001). *Project Number 1186*, Joshua B. Kardon + Company Structural Engineers, Berkeley, California.

Kardon, J.B. (2002). *Project Number 465,* Joshua B. Kardon + Company Structural Engineers, Berkeley, California.

Kardon, J.B., Schroeder, R.A., Ferrari, A.J. (2003, October). *Ethical Dilemmas of Technical Forensic Practice,* 3rd Forensic Congress, Technical Council on Forensic Engineering, American Society of Civil Engineers, San Diego, California.

Kardon, J.B. (2003). *The Standard of Care of Structural Engineers*, Dissertation, Doctor of Philosophy in Engineering, Civil and Environmental Engineering, University of California, UMI Dissertation Services, Berkeley, California.

Kardon, J.B. (2005). The Concept of "Care" in Engineering. American Society of Civil Engineers, *Journal of Performance of Constructed Facilities*, Vol. 19, No. 3, pp. 256-260.

Klein, G. (1999). *Sources of Power: How People Make Decisions,* MIT Press, Cambridge, Massachusetts.

Kolodner, J.L. (1993). *Case-Based Reasoning*, Morgan Kauffman, San Francisco, California.

Kolodner, J.L., Leake, D.B. (1996). A Tutorial Introduction to Case-Based Reasoning, in *Case-Based Reasoning, Experiences Lessons and Future Directions*, David B. Leake, Ed., MIT Press, Cambridge, Massachusetts.

Kumho Tire Co., Ltd. v. Carmichael, 131 F.3d 1433 (1999).

Leake, D.B. (1996). CBR in Context: The Present and Future, *Case-Based Reasoning, Experiences Lessons and Future Directions*, David B. Leake, Ed., MIT Press, Cambridge, Massachusetts.

Loss, J. (1987). The AEPIC Project: Update, *Journal of Performance of Constructed Facilities*, American Society of Civil Engineers, New York, Vol. 1, No. 1, February, pp. 11-29.

Maher, M.L., Balachandran, M.B., Zhang, D.M. (1995). *Case-Based Reasoning in Design*, Lawrence Erlbaum Associates, New Jersey.

Merriam-Webster. (2004). Webster's Third New International Dictionary of the English Language, Unabridged, Springfield, Massachusetts.

Online Ethics Center for Engineering and Science. (1995). Case Western Reserve University, Cleveland, Ohio, http://onlineethics.org/eng/cases.html.

Rosenman, A., Gero, J.S., Oxman, R.E. (1992). *What's in a Case: The Use of Case Bases, Knowledge Bases, and Databases in Design*, CAAD Futures '91, Proceedings, International Conference on Computer-Aided Architectural Design, Gerhard. N. Schmitt, editor, Vieweg and Sohn, Wiesbaden, Germany, pp. 285-300.

Rossi, P.H., Freeman, H.E. (1979). *Evaluation: A Systematic Approach*, Sage, Newbury Park, California.

Sackman, H. (1975). *Delphi Critique: Expert Opinion, Forecasting, and Group Process*, The RAND Corporation, D. C. Heath and Company, Lexington, Massachusetts.

Tronto, J.C. (1993). *Moral Boundaries: A Political Argument for an Ethic of Care*, Routeledge, New York.

Turoff, M., Hiltz, S.R. (1996). Computer-Based Delphi Processes, in *Gazing Into the Oracle: The Delphi Method and its Application to Social Policy and Public Health*, Michael Adler, Erio Ziglio, Eds., Jessica Kingsley Publishers, London.

USGS. (2012). ShakeMap Scientific Background, U.S. Geological Survey, Retrieved from http://earthquake.usgs.gov/earthquakes/shakemap/background.php.

Ziglio, E. (1996). The Delphi Method and its Contribution to Decision-Making, in *Gazing Into the Oracle: The Delphi Method and its Application to Social Policy and Public Health*, Michael Adler, Erio Ziglio, Eds., Jessica Kingsley Publishers, London.

Chapter 4
Investigations and Reports

4.1. INTRODUCTION

One objective of a forensic engineering investigation is to identify the causes of a failure of a component, a system, or a structure. The failure may be catastrophic if it results in loss of life or property, but not every failure is catastrophic. Investigations and reports of failures large and small are important because they can identify causes of the failure and so aid in allocating responsibility for the failure, and perhaps lead to improvements that could prevent future similar failures or reduce their effect.

The forensic engineer should use valid and reliable methods of investigation and of developing opinions based on the investigation. One should continuously update one's opinions as additional information is uncovered during the investigation. Information uncovered and opinions reached during the investigation may affect the course and scope of the remaining investigation. An expert should not present an opinion as final without completing a thorough and exhaustive investigation. This chapter delineates the steps involved in a "typical" failure investigation and production of a report. It outlines the steps that will likely result in the performance of a thorough investigation and an informative and useful report. Investigation steps are likely to include examinations at the failure site, physical sampling and testing at the failure site or at a laboratory, engineering analyses, and reporting the results of the investigation. The client might request a written report during an investigation or at its conclusion. The report as a standalone document must communicate the steps taken in the investigation and the conclusions reached.

Depending on the agreement between the forensic engineer and the client, a failure investigation may conclude with a description or resolution of the causes of the failure, or may extend to the production of recommendations or actual design and execution of solutions or repairs.

No set of guidelines can inform engineers of all the possible techniques and actions needed for every failure investigation and report. This chapter does not prescribe a rigid formula for conducting forensic investigations or for writing reports; instead, it outlines fundamental elements of a logical, valid, and reliable forensic investigation and report to aid forensic engineers in identifying the causes of a failure.

The outlined steps described in this chapter will assist an investigator in the completion of acompetent investigation and report; however, they are not intended to be exhaustive. The reader is directed to an upcoming publication by the Committee on Forensic Investigations titled, *Guidelines for Failure Investigations*, for more information and guidance.

4.2. INVESTIGATION

Allegedly injured or responsible parties involved in a failure of a portion of the built environment will often retain a forensic engineer to assist in determining the causes of the failure. An integral part of that assistance is an investigation. Forensic engineers performing investigations should specialize in the specific field of the investigation and should use valid and reliable investigation methods. The following sections describe several fundamental steps in an investigation.

4.2.1. Prospective Client Identification and Conflict of Interest Clearance

Forensic assignments typically begin when a party contacts the forensic engineer inquiring as to the engineer's availability to help with a problem the party has encountered. The forensic engineer should limit the initial communication with a prospective client to identification of all interested parties and confirmation that a conflict does not exist. (Please see Section 5.3.3. in the Ethics chapter of these *Guidelines* for information regarding conflicts of interest.) An engineer should only accept a failure investigation assignment after determining that a conflict does not exist and should avoid any discussion of details associated with the potential assignment until such time. Communicating details of the incident before the forensic engineer is retained by the client may compromise the investigation. Such communication may preclude the forensic engineer from effectively participating in the investigation, or may result in the engineer being conflicted out of working for another party to the dispute.

Upon confirmation that a conflict does not exist, information concerning the specific matter can be exchanged between the client and the forensic engineer. Because a percentage of failure investigations involve some sort of litigation, all documented communication, written or electronic—may be discoverable; therefore, any miscommunication that may be incorrectly documented may require explanation if litigation ensues. Documenting oral communication may be worthwhile so as to have a full record of the investigation process and progress for the retaining client, as well as to justify the findings of the investigation for possible future litigation.

4.2.2. The Scope of Assignment

After the forensic engineer accepts the assignment, discussion of details surrounding the incident are appropriate. It is at this stage of the investigation that the engineer and the client can establish an understanding of the scope of the assignment and the focus or initial focus of the investigation.

The investigator should be mindful of time and financial constraints imposed by the client at the start of the investigation and as the investigation progresses. Financial constraints should not compromise the quality of the investigation, however, investigation time and financial constraints are often uncertain at the beginning of the investigation. If the forensic engineer believes the scope of the necessary investigation and analysis cannot be accomplished within the time and money constraints of the client, it may be better for all involved that the forensic engineer not accept the assignment. The validity and reliability of the investigation can be compromised unless sufficient time and resources are brought to bear.

The forensic engineer should propose an agreement for the initial services to the client prior to commencement of any services. The agreement should comply with laws governing such agreements in the state where it applies. It should include the billing rates and reimbursable expenses to be charged to the client, explain how additional or extra fees and expenses will be requested and paid, describe how the agreement may be terminated, and incorporate requirements for the timely payment of invoiced amounts. The forensic engineer may consider not providing services until and unless the client executes the agreement. Continued communications between the investigator and client could minimize conflict over billing. If the client requests additional or extra services, both parties should sign an addendum to any written agreement to document and memorialize the request.

4.2.3. Background Information Review and Relevant Research

Following the initial client communication, the forensic engineer should obtain copies of all relevant print documents or other media. Depending on the nature of investigation, these documents may include plans, calculation sheets, expert reports, past repair or remodeling documents, photographs taken of the site in the past, and loading history of the structure including significant loading such as earthquakes or hurricanes to which the facility may have been subjected. The client may provide documents, or the forensic engineer may otherwise obtain them. These documents will assist in the planning and scheduling of the investigation effort and will provide direction for data gathering, equipment requirements, and length of time required to conduct the fieldwork. The forensic engineer investigator should objectively examine these documents without compromising the final opinions. It is imperative that these documents be carefully and thoroughly reviewed as part of the investigation. The possibility exists that some physical evidence may have been removed or altered at the scene, or certain facts associated with the incident may have been inadvertently omitted without verbal communication or the knowledge of the investigator. This step is important to ensure that a thorough and complete investigation is performed.

Research utilizing additional resources may be useful prior to the initial site visit, or any time thereafter. Internet research should be used with extreme caution and restraint because of the unverifiable and possibly unreliable nature of those sources. At present, published peer-reviewed documents are the most reliable resources for scientific information.

4.2.4. Initial Inspection, Examination, or Review

Failure investigations may involve some kind of litigation. In these cases, once agreements for forensic services are executed, the engineer should perform a preliminary site visit or document review to determine whether initial findings can support the client's position. If the initial opinion of the forensic engineer is that the client's litigation position may be tenuous, the forensic engineer must inform the client and discuss appropriate actions. It may be that the initial review or investigation did not uncover facts or evidence entirely in support of the client's position, but did show the forensic engineer where a truthful, valid, and reliable investigation might take place to provide worthwhile support for the client's argument. It is appropriate at any stage of the investigation process that the forensic engineer informs the client of the impact on the client's position of the evidence as it is developed.

The purpose of the fieldwork in a forensic investigation is to provide physical access to the scene of a failure and to allow direct appraisal of the subject of the investigation. A first-hand evaluation can provide the investigator with valuable information and may be a deciding factor in the successful resolution of a dispute that may arise from the failure.

Systematic and methodical field investigations provide the data and evidence to assist the engineering analysis and any laboratory study to identify the causes of failure. Field investigation allows the forensic engineer to see and appreciate the physical conditions. That appreciation leads the investigator to decide what course of technical analyses should be pursued and helps the investigator explain those opinions in a comprehensible manner

An extensive or complex incident may require a preliminary site visit to allow forensic engineers to familiarize themselves with the scope and magnitude of the incident so they can identify any additional resources that may be required for thorough analysis. A senior staff member or other equivalently experienced team member should perform the initial site visit. An incident site or the subject of the investigation may change prior to any follow-up visit; therefore, this preliminary visit should be documented as thoroughly as possible, preferably with extensive photo documentation.

Scheduling and logistics may prevent experts or consultants retained by all interested parties from participating in an initial site visit. In such cases, no part of the investigation site or the subject of interest should be disturbed, or evidence collected, without obtaining appropriate authorization. However, if the site or the subject of the investigation may significantly change prior to inspection by interested parties unable to attend the initial site visit—and thus possibly prevent an equal evaluation of the scene—the investigators present should carry out thorough photo documentation and preservation of the site should be carried out to the to the extent possible. Additional photo documentation should occur throughout the attempted preservation process.

Introductions among parties involved in the investigation and present at any site often takes place prior to the commencement of site work. The engineer should consider creating a written attendance document to record the names, contact information, and party for whom the attendee is working and distributing copies of the document to attending individuals upon request. Often it is not the forensic engineer who performs this task. If attorneys are participating, they often distribute an attendance sheet or a sign-in list. They should also provide the criteria for interaction among the parties in the field. The investigator must remain professional and minimize discussion during the site visit that could be construed as placing responsibility or contributing culpability of the failure on any specific party.

Whether or not a preliminary site visit occurs, the investigators should review all available facts, data, photos, reports, letters, and other documents as part of the initial examination. After conducting the initial investigation and review, investigators should communicate their initial findings to the client and help make decisions concerning the pursuit of the investigation.

4.2.5. Scheduling

Scheduling of a site inspection is dependent on the site conditions, staff and equipment availability, accessibility, time and travel requirements of the specific client, and the assignment type.

Field investigation requirements vary by project. For example, a traffic crash investigation may require crash reconstruction but the investigation may be limited by time constraints and the requirement to reopen the crash scene to normal usage or traffic. In cases where there is a need to return the incident to its pre-incident conditions, the forensic engineer must perform meticulous and thorough documentation of the scene. Still photography and videography should be comprehensive and thorough. In situations involving structural collapse or partial collapse, immediate temporary shoring and stabilization of the structure concurrent with debris removal may be required for safety and to allow recovery, retrieval, and reconstruction. In a failure involving injury or loss of life, rescue and recovery efforts naturally take precedent over investigations. An investigator on site may be called upon to lend special expertise in assisting such efforts. In all cases the investigator should be attentive to site safety as well as supportive of the rescue and recovery efforts.

Many projects are investigated while the subject facilities are in use. This may require coordinating schedules with the occupants to minimize disruption of their normal routine. Occasionally, a catastrophic failure may call for immediate response. In these cases, the investigator may be called to the site on a moment's notice. A camera, tape measure, field notebook, and experience may substitute for more extensive planning and equipment inventory in cases where a fast response does not permit extensive preparation.

4.2.6. Data Gathering and Documentation

The process of gathering information useful for the analysis of any particular project can vary significantly. The investigation of a vehicle crash immediately following the incident may allow only a few hours for data gathering and scene documentation with the entire scene undisturbed, while the investigation of building distress due to soils settlements or expansion can take months to several years.

Evaluation of a vehicle crash that is not performed immediately after the incident can take place months after the evidence is moved and stored at a location away from the accident scene. Structural and soils issues remain at the site, but may be continually changing. For this reason, sketches and notes are useful in conjunction with photographs or video. Sketches to scale are helpful for later review and in discussions with other consultants, but are not always practical under field conditions. These along with notes included can be redrawn to scale in the office. Rough sketches with photos of the sketch area that contain some definitive scale should be relatively easy to reproduce. If available, site or building plans, machinery specifications and other documents prior to site visits can save time during the fieldwork. With advancement of computer and internet technology, an investigator may obtain an aerial map of the project site and may even use them to estimate dimensions of subject buildings. These documents can aid in collecting and preparing more accurate details associated with the incident. Instead of hand sketches, these copies can be utilized at the site to mark the location of data collection.

4.2.7. Photographs and Video

Photography is a required tool in documenting site conditions. It is essential for documentation of patterns, measurements, damages, and scale when objects of known size are included. A good practice is to include a marked scale in the frame of the photographs. Site photographs should include various perspectives and distances. These photographs are the best way to establish the overall layout of a site and enable the investigator to visualize an entire site and better understand and explain the investigation and its context. All specific items or conditions should be documented with a minimum of two photographs, distant and close-up. Varied lighting conditions and camera angles help to assure clear documentation. For subsequent presentation purposes or report production, enhancement of a photograph may be acceptable provided that the caption includes a note explaining what modification or enhancement of the photograph had occurred.

Videography can provide detailed description and documentation of conditions at the scene. Video recorded without audio is acceptable unless the investigator desires to record extemporaneous speech and other sounds.

4.2.8. Interviews

Evidence can be obtained through eyewitness interviews. This form of evidence collection can provide some data that may be unavailable through other sources. The information obtained through this means should be carefully evaluated and cautiously accepted. The statements of eyewitnesses may be useful to the study and patterns of repetitive damages or occurrences.

Recording of the interview is recommended and usually requires permission from the party being interviewed. Whenever feasible, investigators should conduct interviews as soon as possible after the subject incident occurred. Over time, memory of details becomes unclear or is lost. Whenever possible, distribute fact-finding questionnaires to witnesses and collect the responses prior to the fieldwork. Studying responses in questionnaires can be useful during the investigation. However, those responses should only be used as a guide not as proven facts. Only a proper investigation performed by trained investigators should serve as the basis for findings.

4.3. EQUIPMENT SELECTION AND AVAILABILITY

Forensic firms may develop inventories of useful equipment for their normal project requirements. These inventories will tend to grow as the diversity and experience of the groups expand. Typical equipment used for projects might include still cameras, video recorders, metal detectors, construction levels, manometers, distance measurement devices, x-ray equipment, ground penetration radar, infrared camera, ultrasonic impact echo units, moisture and chemical detectors, surveying equipment, power tools, temperature and pressure monitors, and other special equipment. Only experienced personnel familiar with the scope of the investigation should select the essential equipment for it. Many times the equipment will be case specific and possibly cost prohibitive. A reputable supplier is the best source for any unique apparatus or equipment deemed necessary. If the item is to be rented, the forensic engineer should confirm that the equipment was maintained and recently calibrated.

4.3.1. Equipment Usage and Calibration

Equipment must be well maintained, cleaned after each use, and calibrated at the manufacturer's specified intervals. Most equipment can be calibrated by the equipment operator utilizing the manufacturer's documentation. More sensitive tools may require calibration by the manufacturer or a firm specializing in the service of that particular equipment. The investigator should permanently record any such services for future reference and possible legal proceedings. Additionally, keeping the manufacturer's documents and user's manual for each piece of equipment for reference is prudent, since these items may also become exhibits in similar proceedings.

Check all equipment prior to using it in the field. Documentation of these checks and the relative accuracy of the specific equipment is also recommended. For some specialized and calibration-sensitive equipment, documentation of these checks should be maintained. Calibration for accuracy can be determined by measuring known quantities. For example, a metal detector, which measures location and depth of concrete reinforcing, can be verified for accuracy by chipping away the concrete to expose the steel. Only properly trained personnel should use equipment to collect important data.

4.4. SAMPLING AND COLLECTION OF EVIDENCE

Forensic sampling is dependent on project quantities. The number and manner of data sampling is the subject of statistical analysis and evaluation and may become a significant factor in the justification of an opinion. This committee report does not intend to address the subject of statistics, however, the following are examples of sampling approaches:

- Sampling for the failure of a single steel or wood beam might include a failed member and a similar intact member.
- Sampling for identifying concrete strength in concrete elements for which the design strength is unknown might require a minimum of six cores/tests for each floor level, 400 cubic yards of concrete, or 10,000 square feet of surface area, whichever requires the most frequent testing (ASCE/SEI 2006).
- Studies of distressed structures might require the sampling of as-built construction in 20% of the units in a small development, but only 10% in a large development.
- Investigation of a project constructed in phases, by different contractors or subcontractors, will require sampling in each phase for varying or similar defects.

To ensure integrity of the evidence, the forensic engineer should 1) carefully collect and document evidence and physical samples, 2) photograph all collected items prior to movement and packaging, 3) include within the photograph some scale to indicate the relative size of the item as well as an identifiable, existing or applied marking to indicate or confirm its location at the time it was removed, 4) take numerous photographs throughout the collection process to record any changes to the item, and 5) document any damage sustained by the item during movement and packaging.

To satisfy the court and to withstand vigorous cross-examination, the chain of custody of the evidence must include full documentation of when, where, and by whom the evidence was obtained, who had access to it, who stored it, and where it was stored until its arrival in the courtroom. Organization and protection of the evidence and any associated data is critical to the analysis when the incident is subject to any legal action.

Many standards relating to the interaction of evidence have been established by the American Society for Testing and Materials (ASTM) and other professional organizations such as ASCE. The forensic engineer should use these standards throughout the investigation process. These standards included but are not limited to:

- ASTM E62011 Standard Practice for Reporting Opinions of Scientific or Technical Experts
- ASTM E67807 Standard Practice for Evaluation of Scientific or Technical Data
- ASTM E86007 Standard Practice for Examining and Preparing Items That Are or May Become Involved In Criminal or Civil Litigation
- ASTM E118805 Standard Practice for Collection and Preservation of Information and Physical Items by a Technical Investigator
- ASTM E145992(2005) Standard Guide for Physical Evidence Labeling and Related Documentation
- ASTM E149211 Standard Practice for Receiving, Documenting, Storing, and Retrieving Evidence in a Forensic Science Laboratory
- ASTM E233204 Standard Practice for Investigation and Analysis of Physical Component Failures

4.4.1. Organization and Format of Data

The organization and protection of the data collected is critical to the analysis, the office work, and—when legal action occurs—to the mediation and trial process. Field data can be organized by numerous methods. The key to effective organization is simplicity. Because these data may be reviewed by many different parties with various backgrounds and from various disciplines, the organization of the data should allow all involved parties to access and understand it.

4.4.2. Avoidance of Preconceived Notions

Forensic studies and field investigations should gather and record data as found at the site. Every effort should be made to eliminate trying to force the evidence, data, and analysis to fit the consultant or client's preconceived hypotheses. The forensic engineer should not approach data gathering with the question in mind, "How can I find evidence to support my sense that...?" but rather, "How can I find all of the important evidence?" When the forensic engineer has been hired by one party to a dispute, one check for bias involves asking, "How would I do this if I were working for a different party, or for the public at large?" Answering that question could be helpful not only in identifying one's own bias, but also in anticipating the investigative and analytical steps other parties' forensic experts may be considering.

4.4.3. Storage of Samples and Protection of Evidence

Collected samples should be stored in a method that maintains them in a condition closest to the way they were found. Protecting ferrous metals can prevent additional rusting, failed surfaces from rusting, oxidation, impact, crushing, and abrasions. Larger sample storage should provide support for the sample to maintain the shape or configuration existing at the time of field sampling.

4.5. PRELIMINARY FINDINGS AND CLIENT CONSULTATION

After the initial inspection and review, the forensic engineer should consult with the client and present the initial findings of the investigation. If possible, the forensic engineer can propose a preliminary hypothesis of the failure mechanism or cause based on the initial review. This will aid the client's decision regarding the next steps of the investigation, and allow the client to appraise the initially anticipated costs and to evaluate the client's possible legal or financial exposure. At the conclusion of this consultation, the forensic engineer and the client should have arrived at an agreement about whether to continue with the investigation or terminate the process. If the investigation is to continue, the forensic engineer may wish to provide a supplemental proposal for the continued services. If the investigation is to be halted at this initial stage, a written report may not be required.

4.6 ADDITIONAL INVESTIGATION LEADING TO IDENTIFICATION OF CAUSES OF THE FAILURE

Additional forensic examination or analysis along with possible further research will lead to a follow-up consultation with the client. These steps will be reiterated until the client is satisfied or terminates the investigation.

4.7. INVESTIGATION GROUP

There may be circumstances where an investigation is carried out by an investigation group rather than by an individual forensic engineer. A professional forensic engineering group should include a variety of individuals that have backgrounds in areas relevant to the specific subject of the investigation and should include individuals with diverse experience. Considerations in the selection of the field investigation personnel should include case-specific requirements for professional expertise of the individuals and of the group as a whole. In addition to senior staff, the investigative team should include less experienced technical members. Aside from their time being charged to the client at a lower rate than the senior staff, the participation of less experienced members provides valuable input and diversity to the group. Less experienced participants might raise critical fundamental issues that may have been overlooked as a matter of routine or may have been thought to be unimportant by more experienced investigators.

Investigations may require other specialized outside consultants or personnel depending on the complexity and details of the failure being investigated. These team associates can provide specialized expertise as well as guidance on various aspects of the investigation.

4.8. INTERACTION WITH THE MEDIA

Discussions or interviews with media representatives are very sensitive in the field of forensic studies. Many forensic projects involve a legal dispute. Published or broadcast comments by consultants are likely to be taken out of context or abbreviated, and may not accurately represent the statements made or the opinions offered. These published accounts can be used to discredit the consultant. The client, governmental bodies, or attorney should handle all media interaction, unless the client requests or authorizes the forensic engineer to do so. The client might instruct the forensic engineer to direct inquiries by phone or in person from radio, television, or newspaper reporters to the client or the client's appointed representative. Absent such instructions, the forensic engineer should decline to comment to media or other parties. No information should be released unless requested and approved in advance by the client.

4.9. SAFETY

Failure sites should be considered dangerous, and all personnel deployed to investigate such a site must be familiar with basic site safety and should be equipped and trained to deal with the hazards they may face. Safety glasses, hardhats, protective shoes, appropriate clothing, safety vests, hazmat suits, and respirators may all be required. In cases that require a hazmat suit or respirator, the user should be fully trained in their safe and proper use.

4.10. LABORATORY INVESTIGATION

Laboratory investigations may follow the field investigations. An acceptable laboratory investigation requires selection, performance, and documentation of applicable test methods, use of engineering standards, adherence to standard laboratory reporting, and attention to the chain of custody. Occasionally, parties to a dispute will agree to a single set of tests with all interested parties invited to observe these tests.

Once the samples arrive there, the laboratory tests the specimens. These tests must follow approved and accepted procedures, often those established by ASTM. If there is not a standard procedure, one must determine if the testing procedure to be used is valid and reliable, and is consistent with other previous testing programs investigating similar properties. When the testing is complete, the laboratory will generate a report. Where the test standard mandates one, the report must follow the predetermined format. Also, one must be careful to adhere to the agreed-upon chain of custody for the samples throughout the laboratory testing.

4.10.1. Laboratory Reports

Laboratory reports normally contain documentation of the investigation in each of the following categories:

4.10.1.1. Initial Information

- Testing Laboratory
- Testing Technician and all observers
- Date Tested
- Time Tested
- Test Number

4.10.1.2. Test Information

- Test Type
- Test Standard Followed
- Length of Test
- Test Conditions

4.10.1.3. Testing Equipment Information

- Equipment type
- Calibration Record

4.10.1.4. Sample/Specimen Information

- Sample/Specimen Description
- Sample/Specimen Location
- Sample/Specimen Dimensions
- Sample/Specimen Weight
- Sample/Specimen Storage Prior to Testing
- Sample/Specimen Sketch
- Sample/Specimen Photo
- Sample/Specimen Alterations/Additions (if any)
- Date Obtained
- Sample/Specimen Number
- Sample/Specimen Chain of Custody (described)

4.10.1.5. Test Results

- Test Result
- Plots/Graphs/Curves
- Test Analysis Followed

4.11. OFFICE INVESTIGATION

The axiom, "You can never have too much information" best describes the goal to assemble all available documents and other information when researching the history or origin of a failure or potential failure. When gathering documents and information in addition to the original design drawings and construction-era documents, the investigator should continue to assemble and organize relevant documentation until the investigation is closed.

Tracking down sources and available documents is often exhilarating as well as frustrating. Failures may occur long after the original designers and builders of a facility ended their involvement with the facility owner, or the owner may have archived or destroyed many documents describing the design, construction, and history of the project. This makes the search more difficult and time consuming. However, once elusive and important documents are found and assembled, the investigator can move forward with greater self-assurance. For many, the hunt for and discovery of historical documents is an exhilarating and gratifying experience.

4.11.1. Data Gathering Phase

Data gathering commences with the first notes taken during a telephone call or visit from the client. Even though there may be many people involved, the current owner should be the focal point and origination of the investigation. If they are capable, the facility owner's engineering staff should conduct the initial research and data gathering since the owner should have design documents contained in the project files. As a minimum, the owner should assemble the layout drawings describing the structure and its location. From this beginning, the investigator can reconstruct and visualize the design and history of the original project. The investigator must be receptive to ideas and concepts presented without preconceived notions. There will be many roadblocks (access to owner's files, uncooperative subjects, etc.) during the data gathering that may hamper the investigation, but an attainable goal and drive to reach that goal will usually prevail.

4.11.1.1. Construction Documents

Starting with a single design drawing (or other form of document), the investigator may discover other leads to follow. The goal is to assemble a comprehensive package of information about the subject structure. Categories of such documents include:

Original Drawings and Calculations: Original calculations may represent the "Holy Grail" for the investigator because they may disclose the intent, methods, and choices of the original designer of the facility. It is more probable, however, that design drawings may be the only documentation to be uncovered during an investigation. Design drawings often lead to many other referenced items that will aid the investigator. From the title block alone, the name of the Architect/Engineer (A/E) who designed the project can be obtained, possibly with an address and telephone number. Initials or names along with the drawing signature dates may be shown in the title blocks, but the most important identifier is often the project number. The project number is generally a unique identifier assigned to a project by the owner or by the designers and their consultants. Project numbers can lead to additional files contained in the owner's or A/E's document storage that will aid in the investigation. Also contained on many drawings are reference drawings, design criteria, and other notes, which further expand the investigator's search and data gathering. Calculations can enable the investigator to match the analyzed structure with the design drawings and "as-builts" while presenting design criteria, design methodology, and individual structural analyses, so that all analyses do not have to be recreated.

Geotechnical Reports: For many sites, foundation investigations are conducted prior to beginning design work. Foundation investigations are commissioned by the owner, and copies are usually available in the owner's files. In addition to subsurface explorations and foundation recommendations, site-specific seismic time histories or acceleration coefficients may be included in a geotechnical report. As a minimum, the geotechnical report includes boring logs (soil classification) and boring locations of the existing site soils. Recommended soil bearing capacity, types of foundations, and anticipated settlements are also usually included in the report.

Specifications: Specifications are valuable because they may contain project- or site-specific conditions and construction procedures that may not appear in any other document.

Inspection Reports: Structural observation reports prepared by engineers or building officials may lead to critical information regarding deficiencies that might have been noticed during the original construction.

Test Data: Data from field and laboratory tests performed during the project construction may provide valuable information about the materials used in construction.

Miscellaneous Documents: Correspondence, change orders, contracts, field diaries, daily reports, requests for information (RFIs), architectural supplemental information (ASIs), applications for payment, and memoranda.

4.11.1.2. Applicable Codes

Review available drawings, specifications, and calculations for notes referencing design codes, material codes, and other design guides. Verify that any codes referenced were those applicable at the time of the design and construction of the subject facility. If no codes are referenced, determine the applicable code at the time of design based on the date of the design documents.

4.11.1.3. Construction History

Documents retained by the owner or contractor during construction may provide insight into the difficulties encountered, and may provide tests and reports required by the design documents. Construction photographs, report logs, and engineering sketches detail the time-line history of the project while providing a record of the "as-built" situations that may not be reflected in the original design drawings. The following items are generated during many projects, whether large or small:

- Soil Testing
- Bolt Tightening Tests
- Pile Driving Logs
- Concrete Compression Tests
- Steel Mill Test Reports
- Construction Field Sketches
- Inspection Reports
- Photographs

During the lifetime of a structure, the owners may change or modify the structure many times. Some of the renovation projects will be accomplished with proper engineering practices, but many projects are initiated and completed by local on-site staff and no documentation exists. The investigator must review site-retained files for any modifications and compare them with the actual site investigation results. Tracking down the actual documentation is time consuming and may not be necessary if sufficient information is gathered during the field investigation.

4.11.1.4. Maintenance History

Maintenance records may provide another valuable resource. For example, reports of a persistent roof leak may explain corrosion or rot. An extended power outage may explain water damage where a sump pump was unable to operate during a period of heavy rain and flooding.

4.11.1.5. Shop Drawings

Shop drawings are produced by the contractor or subcontractors to aid in the fabrication or erection of portions of the built facility, and are often required by the design or contract documents. Shop drawings may consist of structural steel fabrication drawings, concrete reinforcement fabrication or placement drawings, ductwork fabrication drawings, piling detail drawings, and the like. Shop drawings will differ from design drawing because the original design drawings do not show complete information of every piece or connection. The investigator should obtain shop drawings and compare them to field conditions and design drawings. This is often time consuming and may not be efficient, but shop drawings may better depict the actual field conditions during construction than will the original design drawings.

4.11.2. Data Assembly

The final step in office work is to compile and document the prepared materials including sources of information, codes, standards, drawings, references, and assumptions.

4.12. REPORTS

The report describes the services performed by the forensic engineer, presents conclusions drawn by the engineer, and contains detailed information supporting the engineer's conclusions. The written report may be the only product of the service provided to a client in consideration of its payment of fees to the investigator. A well written report should be factual and dispassionate, and should be comprehensible and grammatically correct. This section describes several parts of a report that the Committee views as important or relevant to include or consider.

The report must be clear and orderly and should be carefully developed to include all information used to substantiate the conclusions. Appendices attached to the report should include the data gathered to support the conclusions. If these data are included, they must clearly support the conclusions and should be organized so that the reader can understand the logic used to reach the conclusions.

Prior to the final issuance of the report, a draft should be reviewed and edited for technical and grammatical accuracy and completeness. Although the body of the report will likely contain technical wording, the summary and conclusion should be written concisely and clearly so that it can be reasonably understood by the layperson.

Where an investigation is part of a legal dispute, the forensic engineer's work product may be subject to special rules of evidence. The forensic engineer should confer with the client and client's counsel as to the correct procedure for reporting interim results as well as final conclusions. See Chapter 6 of the *Guidelines*.

This section on reporting is based upon the assumptions that the report is intended for an external client and that it may be presented in a legal arena. Many of the basic principles of good reporting, however, are equally valid for internal clients or for investigations where legal proceedings are not anticipated.

4.12.1. Purpose of the Report

Regardless of the thoroughness of any investigation or clarity of findings, if the results are not clearly communicated, the hard work carried out as part of the investigation may be lost. The report is often the only opportunity to present the findings of the investigation. The expert may be limited in any testimony to the information and conclusions presented in the report. If details are not included, if the argument presented is unsound or not intelligible, or if conclusions are inadequately presented, there may not be an opportunity to remedy these errors. Preparing and presenting a report is therefore a vital part of any forensic engineering investigation. The investigation report may have serious financial and legal consequences to the forensic engineer's client and others. It is therefore essential that the report is clear, well reasoned, and concise.

A report should read logically from beginning to end, observations should precede analyses, and opinions will form naturally in the reader's mind when facts are explained in a straightforward way. It is most effective to present the facts and findings of a report first, with no editorializing. Later, in a discussion section, the report writer can synthesize the elements of the investigation with findings and present their importance. This approach forces a discipline that typically enhances the work presented.

Regardless of the choice of style, it is important to communicate findings in a way that is easily understood by a layperson. The more accessible the language, the more effective the report.

4.12.2. Basic Reporting Issues

Basic issues in any report include provision of adequate documentation of statements, references, and sources of information, and adequate definition of the various parties involved. In all instances, the report should be accurate and clearly differentiate fact from personal observation and professional opinion. Even though the use or objective of the report may be to help identify responsible parties and potential liabilities, professional ethics must be maintained in the written word.

It is important to clearly define terms that may be ambiguous or simply not understood. Terms such as "good," "bad," "adequate" can be interpreted in multiple ways. Floor levels in a building such as "cellar," "ground," or even first, second, third, etc., are best defined, either through a drawing reference or by description. Clearly define technical terms in the body of the report. Those most interested in the findings of the investigation are rarely technically knowledgeable. Provide clear definitions that are unambiguous and easily understood.

With regard to attributing information to others' statements, take care to clearly differentiate a direct quote from a paraphrased statement. Whether quoted or not, it needs to be clear in the report if statements were from depositions, witness interviews or declarations, other reports or were heard directly by the report writer. When possible, provide the source, date, and reference location. Where witness statements are quoted, identify the interviewer, including name and affiliation. Where appropriate, name the party retained by the interviewer (i.e., the client or agency). Use of standard footnote citation and bibliography protocols will usually suffice, though clients—particularly attorneys—may have particular preferences with regard to the method of giving attribution.

Use industry-accepted references whenever possible to substantiate statements. Although there may be circumstances where test procedures need to be developed for a specific problem, where possible, reference test protocols to acceptable sources. For example, ASTM test standards are prepared under ANSI protocols, as are ASCE standards. The hallmarks of the American National Standards process include (ANSI 2012):

- Consensus on a proposed standard by a group or "consensus body" that includes representatives from materially affected and interested parties.
- Broad-based public review and comment on draft standards.
- Consideration of and response to comments submitted by voting members of the relevant consensus body and by public review commenters.
- Incorporation of approved changes into a draft standard.

- The right to appeal by any participant that believes that due process principles were not sufficiently respected during the standards development in accordance with the ANSI-accredited procedures of the standards developer.

ASTM test standards typically contain statements regarding precision and bias in the test results. Tests are designed to be repeatable, such that independent researchers will obtain the same results following the stated protocols. Any tests or investigations that are referenced within the report should be repeatable, to the extent that the information or test materials upon which the report is based are still available.

4.12.3. Report Organization

A report contains five basic elements: introductory information, description of the issues, information supplied, investigation results, and conclusions. The introductory information and description of the issues typically precede the investigation results and conclusions, separated by information supplied. The following represents the standard sections that may be contained in a report prepared as part of a forensic engineering investigation (the title page and executive summary are considered to be outside of the main report in this list). The order is only a suggestion, as the style, complexity of issues, and volume of material will all affect the final layout of the report.

- Cover Page
- Title Page
- Table of Contents
- Executive Summary
- Introduction / Background
- Scope of Work
- Organization Chart
- Approach
- List of Documents
- Other Information (supplied or obtained)
- Site Visits
- Description of the Site/Building
- Review of Documents
- Site Observations
- Analyses
- Test results
- External Reports
- Discussion
- Conclusions
- Recommendations
- Disclaimer
- Signature

- Photographs, Charts, Graphs and Figures (if not included in the body of the report)
- Appendices

These elements may not all be necessary in any given report, such as External Reports, Test Results or Organization Chart. Some sections may be combined into a single entry, such as Introduction and Scope of Work. Additional sections may also be added as required. These sections are discussed more fully below.

4.12.3.1 Cover Page

In addition to the report title, the report should contain a unique identifier. Appropriately identify interim reports and drafts as such and date them individually. Include this information on every page of the report within the header or footer.

4.12.3.2. Title Page

In addition to the name and affiliation of the writer, include project reference numbers, the project name, and to whom the report is addressed. Add limitations as to distribution and use. Information on the cover page may also be duplicated here.

4.12.3.3. Table of Contents

A table of contents is important for the reader for navigating through the report sections, photographs, figures, and other data. For very large reports, the table of contents may include more than one level of subsections to assist in rapidly locating information.

4.12.3.4. Executive Summary

Because many forensic engineering reports may be lengthy and technical, it is necessary to present a concise summary of the entire report for rapid review by key decision holders. The executive summary typically provides a statement as to the issues, scope of the investigation, brief statement regarding limitations of the investigation and report, and the main conclusions. The executive summary omits details, photographs, drawings, tables, and other background or supporting information and documents. The executive summary should be limited to one typed page, though it may sometimes be longer. It should contain no information contained that is not also contained in the body of the report.

4.12.3.5. Introduction and Background

At the beginning of any report, it is important to indicate who the client is and, broadly, what the issues are. This will include a brief description of the scope of work that is addressed in the report and identify the project location. With regard to the scope of work of the engineering investigation, this should not deviate from the signed proposal for the project. Key individuals may be named for those investigations involving more than one person.

Depending upon the level of complexity, the report may include a separate section describing the issues. This is typical of investigations focusing on a list of allegations, often listed in a complaint or plaintiff's expert report. Where no such predetermined list exists, the issues may be simply stated as to determine the cause of collapse.

4.12.3.6. Scope of Work

This section describes the basic issues and objectives in the scope of work, including the activities carried out and any limitations or expansion of the agreed scope. The report writer may include in this section a copy of the agreement between the forensic engineer and the client.

4.12.3.7. Organization Chart

Where multiple individuals or outside consultants or test laboratories contribute to the investigation or lead individual tasks, it is often helpful to include an organization chart showing the roles and relationships of these various individuals and groups, along with the role each individual or organization plays.

4.12.3.8. Approach

For complex projects involving multiple tasks, a description of the approach to the investigation may be beneficial in aiding the reader to understand how these different elements have been used towards reaching the final conclusions. However the investigation is undertaken, it should be readily understandable to both the client and other interested parties. A description of the approach is best included where it is also found necessary to include an organization chart.

4.12.3.9. List of Documents

The report should provide a list of the documents received or reviewed. This also assists in establishing the limitations of the investigation. Missing documents or those that have not been reviewed can then be readily identified. Additional documents can form the basis for supplementary reports and even changes to the conclusions.

4.12.3.10. Other Information Supplied or Obtained

There may be additional information on weather records, past earthquakes, material properties and behavior, or a structure's history. In these cases, depending upon the complexity of the issues, the extent of information, and the other details specific to each project, separate sections for each of these efforts may render the report clearer to the reader.

4.12.3.11. Site Visits

It is good practice to provide a separate section listing site visits. Site visits can include both visits to the project site as well as to other locations where physical evidence has been stored. In some instances, the investigator might add a list of visits to others offices or premises, such as laboratories, manufacturers, or information repositories.

4.12.3.12. Description of the Site or Building

Forensic engineering investigations typically are focused on a physical issue having a specific location. Where processes are under investigation, multiple locations may be affected. In either case, a description of the building or site or other items that represent the location or focus of the investigation needs to be described.

Typically, for buildings, this will include the following items:

- Physical address.
- Age of the building (may be more than one where extensions, renovations or other modifications have occurred in the past).
- Basic physical description: number of floors, roof type, existence of any basement or crawl space, construction type (precast concrete, steel frame, etc.), and occupancy classification.
- Where relevant, additional physical information such as number of apartments, description of retail, office or other space, dimensions, number of parking spaces, etc., may be provided.
- The names of the developer, designer, architect, engineer, contractor, and other relevant parties, where known.
- Other pertinent details, such as intersecting roads, neighboring structures, and orientation.

4.12.3.13. Review of Documents

For many forensic projects, this section may represent the entire investigation. Often, physical evidence no longer exists and mathematical analyses are not required. The investigation then depends upon the careful analysis of the various documents that remain the sole source of information regarding the client's concerns.

It is best to create a methodical approach to the analysis of the documents, whether by author, issue, or date of creation. Regardless of the method, a full description of each document, including the date of creation, type of document, title (if any), subject line (if any), author, and the recipient all should appear in the reference. Where the analysis may include multiple emails, more than one may be sent to or from the same individuals in any given day, therefore, the time of transmission or receipt will help to fully identify the document.

Reference Bates numbers where they are present, as these also provide an easy method for all parties to locate documents at a later date. Bates numbers are individual identifiers placed on each page of a document. Originally imprinted on each sheet by the Bates Automatic Numbering-Machine, a stamp originally manufactured by the Bates Manufacturing Company, Bates numbers these days are often electronically produced. They typically include initials identifying the originator of the document and a unique page number.

4.12.3.14. Site Observations

In the case of site visits, an investigator will have typically made observations pertinent to the investigation as well as measurements, site tests, sampling and other activities. The report should describe the details of these observations. It is best if any testing or probing follow industry accepted protocols, such as those created by an ANSI standards-accredited organization. As discussed more fully below, this will greatly assist in having results that are acceptable to the client, opposing parties, and the court.

Photographs are most likely to be referenced in this section. Photograph references should include a method to identify location, date, and source of the photograph. Pertinent details may be highlighted by cropping, enlarging, circling, or other means that does not alter or misrepresent the image.

Where comparative photographs are used, it is essential to differentiate these in the discussion.

Some observations may not be relevant to the immediate objectives of the investigation. When such observations have no bearing on allegations, defenses, or other arguments presented elsewhere (including those made by others in their respective reports), it is permissible to omit comments regarding those observations. Where ethical considerations would otherwise require comments to be included, such as those related to immediate safety issues, the investigator can communicate those observations and comments to the client in a separate document.

4.12.3.15. Analyses

A feature of many forensic engineering investigations is analysis of data or situations, whether real or hypothetical. These may be mathematical or graphical. Mathematical analyses include, for example, simple static analyses of structural components, computer modeling and analysis of structures subjected to dynamic or static loads, blast analysis, wind vortex shedding calculations, footfall analysis, and computational fluid dynamics. Graphical analyses include solid modeling of a debris field, accessibility and egress issues, and computer-based analysis of constructability issues using building information modeling (BIM).

For all analyses, industry accepted software or other industry accepted techniques should be used. Provide proper reference to the source of the software and validation procedures used to prove the software.

A description of any analysis, including objectives, modeling and methodology is best included in introductory statements regarding the analysis results to be discussed. In addition to analysis results, include sensitivity studies on the effect of assumptions made in the analysis, where available. A list of the basic assumptions used in the analysis is important to enable the limitations of the computations or graphics to be clearly identified.

4.12.3.16. Test Results

As for the analyses, provide proper reference to the source of the test protocols including modeling methodology, equipment used, and objectives. As discussed at the beginning of this chapter, it is best if the test procedures used follow industry-accepted protocols, such as those created by an ANSI standards-accredited organization. This will also contribute to acceptable results.

Where such test standards are used, reference to the standard, including the year of publication or approval, is necessary because standards are regularly revised. Where such test standards are not used, for whatever reason, it is preferable to include the full test protocols that were implemented. This may include explanatory details concerning the protocols, assumptions upon which they are based, and limitations of their use. Sensitivity studies on the effects of varying those assumptions, where time and cost permit, can also be very helpful.

4.12.3.17. External Reports

The use of external consultants or external test laboratories typically results in reports prepared by those individuals responsible for the execution of the work in those outside agencies. The results of these reports will then be incorporated into the main report, whether as data for use in analysis, data for testing, or as supplemental information for developing the arguments in the discussion. As for all external sources of information, the forensic engineer should provide a complete reference to the source of that information.

Such external reports are most often generated through the use of subcontracts from the forensic engineer or the engineering company to the individuals or companies conducting the external work. It is best to include as part of any appendices copies of those contracts as well as copies of the external reports that were generated. This should provide all needed information regarding the professional relationships and scope of work requested.

4.12.3.18. Discussion

This is the main focus of the report. It is within the discussion that the various facts, observations, references, analyses, and any other data are pulled together. New information should not be included; rather, the discussion synthesizes what has already been presented. If these items are not discussed, there will be little or no reason for them to have been included in the foregoing sections of the report.

It is acceptable to use outside references within the discussion. If outside references are used to bolster an argument, provide further clarification to a difficult concept, demonstrate the validity of an approach, or for any other reason, include a full citation to that reference. These references should be included with the earlier list of documents.

The discussion should be logical, based on factual information (see the earlier discussion on basics, section 4.12.2.), and clear in its structure. As for the entire report, if the discussion does not clearly communicate the method by which conclusions are deduced, the hard work carried out as part of the investigation may be lost. The discussion is the primary opportunity to show the relevance of the data just presented and to justify the conclusions based on that data. No matter how the discussion is presented, the information and its interpretation should be consistent. Where inconsistencies are identified, the engineer should explain them.

If the argument presented is unsound or not intelligible, testimony concerning the report's conclusions may be inadmissible in court.

4.12.3.19. Conclusions

The conclusions summarize the results of the entire investigation as presented in the report and should satisfy the initial objectives. Where the objectives cannot be satisfied, summarize the reasons previously analyzed in the discussion. Conclusions must be based upon the information presented. It is inappropriate to present any new information or discussion at this point. Conclusions must be supported by a well-reasoned set of logical relationships based upon the facts presented in the report.

4.12.3.20. Recommendations

Investigations often require further effort, whether due to the receipt of additional information, a need for additional analyses or further testing, or other activity beyond the forensic engineer's initial scope of work identified during the investigation. Further effort may also be recommended due to limitations of the work that arise during the work. For whatever reason, if appropriate and acceptable to the client, interim reports may present recommendations for further investigation.

If appropriate and acceptable to the client, a final report unrelated to a continuance of the investigation may present recommendations on, for example, repair schemes or safety issues. In some instances, recommendations may be required, such as when cost estimates are also being prepared based upon recommended repair schemes.

4.12.3.21. Disclaimers and Limitations

It is typical to include a general disclaimer in any forensic engineering report, as the investigation will always be limited as described above. Although a report may be "final," additional information may come in from other sources for many reasons, including discovery proceedings, receipt of others' reports, or additional investigative activities. The forensic engineer must therefore reserve the right to amend the report to the extent dictated by the new information.

Additionally, if not stated elsewhere, state the limitations on the investigation activities themselves. These may include limitations due to funding, time, access, general scope restrictions, or any other reason. Generally, these will already have appeared in the section addressing the scope of work of the investigation.

The report should describe how it can be used, if that is not already described in the scope of work, should be included. The *National Practice Guidelines for the Preparation of Structural Engineering Reports for Buildings*, (CASE 2012), published by the Council of American Structural Engineers, contains several possible disclaimer statements, in addition to information on other report types.

4.12.3.22. Signature

The report should provide the name, affiliation, and title of those taking technical responsibility for the preparation and contents of the report. This may be done either at the beginning of the report on the title page or after the disclaimer. The report should bear professional engineering seals and signatures as required by law.

4.12.3.23. Photographs, Charts, Graphs, and Figures

Where not included in the main body of the report, separate sections at the end of the report may contain photographs, charts, graphs, and figures.

4.12.3.24. Appendices

Additional appendices may include other useful data such as photographs, hand sketches, field notes, miscellaneous correspondence, detailed test data, calculations and the qualifications of the individuals responsible for the investigation. While it can be convenient to place a copy of all referenced documents in the appendices, thus providing a single information source for later reference, the size and volume and ensuing difficulties with distribution and storage of attempting this may be prohibitive.

4.13. CONCLUSION

The task of performing a forensic investigation can be daunting and difficult; however, at times it is the most intellectually challenging endeavor in the field of engineering. The forensic engineer is often required to visually examine an object, component, or system that was physically damaged or mechanically compromised by natural or manmade effects. The extent of damage of the object, component, or system can range from a simple fracture or failure of a small component utilized in a medical device prototype which may result in injury to a single patient, to catastrophic collapse of a structure resulting in the loss of many lives. The forensic engineer's task of helping discover and describe the cause of the incident must be driven by ethical standards as exemplified in the ASCE's Code of Ethics.

The reputation of the forensic engineer rests on the care and judgment of the forensic engineer, and on the practicability of the forensic engineer's recommendations. Forensic engineers have a responsibility to meet the standard of care of forensic engineers in providing their services. A principal means of showing that level of care has been maintained is the creation and preservation of accurate records of the forensic engineer's investigation, and the production of a clear and responsive report.

4.14. REFERENCES

ANSI Essential Requirements: Due Process Requirements for American National Standards,(2012), American National Standards Institute (ANSI), Washington, DC.

ASCE/SEI 41-06, Seismic Rehabilitation of Existing Buildings, (2006) American Society of Civil Engineers, Reston, VA.

ASTM International, West Conshohocken, PA.

Council of American Structural Engineers (CASE) and American Consulting Engineers Council (ACEC), (2012), A Guideline Addressing the Preparation of Structural Engineering Reports for Buildings, CASE 962-A, Washington, DC.

Chapter 5
Ethics

5.1. INTRODUCTION

This chapter on ethics could—if brevity were its only goal—begin and end with a simple recitation of the ASCE *Code of Ethics*, including Canon 3.c which specifically applies to expert witness services: "Engineers, when serving as expert witnesses, shall express an engineering opinion only when it is founded upon adequate knowledge of the facts, upon a background of technical competence, and upon honest conviction" (ASCE 2006) However, this simple canon lacks the shades of nuance that typically enter the discussion whenever engineers consider whether their (or others') work has crossed a line from ethical to unethical practice or behavior. Often, the question of whether an action—or even a single answer to a question—falls within the ethical practice of forensic engineering can only really be asked within the context in which the action is taken. For this reason, this chapter will take up the topic of ethics in a range of contexts, starting with situations where there is likely to be little or no debate, and escalating the examination of ethical issues through the dispute resolution process. Topics span the range from taking on an assignment, up to and including providing testimony in the courtroom, where situational ethics is most likely controversial.

The entire ASCE Code of Ethics applies to all ASCE members, including those members providing forensic engineering services. Forensic engineer members of ASCE are subject to the requirements of the entire Code of Ethics, not just Canon 3.c. which specifically addresses members' obligations as expert witnesses The following chapter sections will explore how the Code of Ethics applies to each aspect of the practice of forensic engineering.

Because questions of ethics can be complicated by context, this chapter will introduce a story about one interaction between engineers who found themselves on opposing sides of an investigation in which a legal dispute seemed inevitable. Aspects of this story will be introduced as they pertain to the sections below.

5.2. THE ASCE CODE OF ETHICS

The ASCE Code of Ethics includes Fundamental Principles, Fundamental Canons, and Guidelines to Practice under the Fundamental Canons of Ethics (ASCE 2006). The Appendix of these *Guidelines* contains the full Code, including guidelines to practice. The reader is directed to the Appendix for the complete language of the Code. The following sections address some of the activities of a forensic engineer where ASCE Code of Ethics principles and canons apply.

5.3. SOLICITATION OF WORK, FIRST CONTACT, PROJECT ASSIGNMENT, AND CONTRACT

Ethical considerations apply first when an engineer is contemplating whether to take an assignment. For simplicity, consider that the initial contact has been made by a party with an interest in knowing why something failed. Perhaps it is the owner of a building where a section of structure has deformed excessively or has otherwise performed in an unanticipated or undesired manner. It may be a contractor who was called back because of a problem that has developed in a completed project. Further, although there is no mention at this point that the investigation may lead to a legal dispute, as a practicing forensic engineer, the contacted party should assume it might. Nevertheless, the forensic engineer may be retained initially to explain what happened and why. Perhaps once these questions have been answered the engineer might be asked to participate in the resolution of a dispute or asked how to perform repairs. Design of repairs is not the subject of these *Guidelines*. However, for an investigation that precedes a dispute, the scope of services included in the agreement between the client and the forensic engineer must reflect what is expected in the initial contract. Further, it may be appropriate and ethical for the engineer to lay out in the agreement, or in a separate letter, the expected, anticipated, or typical scope of services, from investigation through testimony or settlement. If this is not done initially, some clients that are unfamiliar with the process may see the increasing scope of engineering services as "bait and switch" on the part of the engineer.

Consider the following example. Engineer "Q" is asked to investigate the collapse of a building under heavy snow. The assignment comes from one of Q's regular clients, in this case, the insurance carrier that covers the building. Q is later approached by a second party. Explaining that he is already involved, Q suggests the second party call a colleague in a different firm, engineer "R," whom he understands has a strong background in this building type. The second party does so, but R states that he would never take a job that might place him on the opposing side against the designers of such a building. In this case, R has revealed a bias. While this may seem like a possible ethical breach, it may simply be a business decision. If R has worked extensively for designers and contractors specializing in buildings of this type, he may not wish to get involved in an investigation where he might find fault in the work of a former client or associate. In and of itself, R's attitude is not unethical so long as it does not extend to the technical aspects of his work.

5.3.1. Solicitation of Work

In considering how ethics applies to solicitation of forensic engineering assignments, it is appropriate to assume, unless notified to the contrary, that a failure investigation may not lead to a lawsuit. However, solicitations by ASCE member engineers, including those offering forensic engineering services, must conform to the Code of Ethics, and may also be proscribed or limited by state statutes or regulations.

State registration boards generally have rules regarding allowable forms of solicitation of work by professional engineers. Codes of ethics for professional associations also address this topic. Generally, advertisements placed in phone books, trade journals, or professional society publications are appropriate if the information is professionally and truthfully presented. Some engineering firms advertise in newspapers and on radio and television. The medium of the solicitation is not as important as the content when ethical aspects are being measured. The subject of advertisement of forensic engineering services is included in Chapter 7, "The Business of Forensic Engineering."

The ethical principle and canon that most closely define what is or is not ethical in advertising relate to honesty, objectivity, and impartiality. In the ASCE Code, Fundamental Principle 2 states that "(E)ngineers uphold and advance the integrity, honor and dignity of the engineering profession by," among other things, "being honest and impartial." Canon 3 states that, "Engineers shall issue public statements only in an objective and truthful manner." By its nature, advertising is a form of biased speech, but anything included in an ad must be true. At a minimum, advertising copy should not offer practices that fall outside the principles and canons of the ASCE Code.

As an example of a state statute governing public statements and solicitation of work by licensed professional engineers, the Commonwealth of Massachusetts (CMR 2012) includes the following:

> *4.04: Public Statements*
>
> *Registrants shall issue public statements only in an objective and truthful manner.*
>
> *Registrants shall be objective and truthful in professional reports, statements or testimony.*
>
> *Registrants may express publicly a professional opinion on technical subjects only when that opinion is founded upon adequate knowledge of the facts and competence in the subject matter.*
>
> *Registrants shall issue no statements, criticisms or arguments on technical matters which are inspired or paid for by interested parties unless the registrants have prefaced their comments by explicitly identifying the interested parties on whose behalf they are speaking and by revealing the existence of any interest the registrants may have in the matters.*

4.06: Solicitation

Registrants shall avoid improper solicitation of professional employment.

Registrants shall not falsify or permit misrepresentation of their own or their associates' academic or professional qualifications. They shall not misrepresent or exaggerate their degree of responsibility in or for the subject matter of prior assignments. Brochures or other presentations incident to the solicitation of employment shall not misrepresent pertinent facts concerning employers, employees, associates, joint ventures or past accomplishments.

Registrants shall not offer, give, solicit or receive, either directly or indirectly, any commission, or gift, or other valuable consideration in order to secure work, and shall not make any political contribution intended to influence the award of a contract.

Texas State law has similar wording. In considering the kinds of rhetoric that clearly fall outside these rules, the offer to perform a "free assessment of their case" may violate the third rule on solicitation because it offers to provide a valuable service (something for which an engineer might typically expect to be paid) in exchange simply for the opportunity to be hired. Further, it is not clear how an evaluation can be made to a degree sufficient to tell a potential client that they may have a case without actually performing the preliminary engineering work necessary for an initial understanding of a failure or problem. By offering to perform for free as much engineering as might be necessary to make that determination implies that a decision about what happened, why, and who is at fault can be made before an adequate investigation has been made. If the investigation begins with a commitment to a specific answer, that investigation clearly represents work that is no longer impartial, objective, or unbiased.

In the example story, R called the building manufacturer, explained his experience with pre-engineered steel buildings, and was retained by the manufacturer to investigate the collapse. This example includes no specific knowledge of what passed between R and his new clients, nor what he promised them. R initiated the contact, and the building manufacturer retained him.

5.3.2 Initial Client Contact

The initial client contact will most frequently occur by telephone. Whether the client calls the engineer, or the engineer approaches the client, the first interaction is important in establishing the ground rules for the engagement. If the potential client already knows that a legal dispute is coming, he may be shopping for legal help as well as technical help. With no knowledge of how an engineer's and an attorney's roles differ, the client may ask up front whether the engineer can prove something. In this case, the engineer is, in effect, being asked whether the engineer is willing to sign on to a case and take a side, even before the engineer knows the facts. It is extremely important to state that engineers cannot be advocates for a client's case, or have an opinion concerning the technical aspects of the dispute, unless and until they perform an appropriately thorough and complete investigation. However, engineers can promise they can and will conduct the appropriate investigation to help determine what happened and why. The engineers will then explain their results in a way that will allow the client and legal counsel to make their best case, or to help them settle their dispute as early as possible if the evidence indicates that is the best they can do. Any response that implies a willingness to adopt a position on the cause of a failure or an attribution of responsibility without adequate investigation or the application of valid and reliable method may mislead the client into thinking that the engineer can be an advocate for the case without an adequate investigation, or independent of the technical evidence, either of which would be unethical.

The forensic engineer should have ethical objectives and principles in mind, as well as business and technical considerations, during any initial contact. The forensic engineer's attitude and tone should convey objectivity, competence, thoroughness, professionalism, financial integrity, and honesty.

If the work is within the forensic engineer's expertise, and the potential assignment meets other requirements of the forensic engineer's business, the engineer will naturally want to provide the services. However, this goal should not control the direction or outcome of the contact. Nor should the discussion focus on ways that the expert could support the client's position, because it would not be possible without an investigation for the engineer to make this conclusion. It certainly is ethical for the expert to assist the client in defining the scope of the investigation, and because the purpose of an expert witness is to assist the trier of fact in answering technical questions where the trier of fact lacks the particular knowledge the expert possesses, it is also ethical and appropriate that the expert witness present as testimony the factual, valid, and reliable technical opinions in the best light for his client. Early in the contact, the engineer should strive to establish whether the topic is in the engineer's field and that the client is willing to authorize a thorough, objective investigation. The client should also be willing to concede that the decision to provide testimony in regard to the findings will occur after the investigation reaches its conclusion. Compensation for the services should be discussed, especially the need to affirm that the client is willing to pay for the investigation even though it may conclude that the client may not prevail in court. Any potential conflicts of interest should be disclosed and a discussion of their impact should occur prior to setting the final terms of the contract.

The forensic engineer should also obtain background information such as:

- What happened to occasion this call?
- How did the caller obtain the engineer's name?
- What are the circumstances surrounding the failure?
- What is the relationship of the caller to the failure?
- What are all the perceived technical issues, and are they totally within the forensic engineer's expertise?
- Does the caller plan to engage other forensic engineers in the same field? In other fields? Should the forensic engineer recommend that other expertise be brought in?
- What level of effort is expected, and what budgetary or schedule limitations is the caller faced with?
- Has the caller considered the terms of engagement?
- How and when will the engagement get started?
- When will preliminary results of the investigation be needed?
- If an attorney makes the call, will the attorney or his client be signing the contract and paying the bills? It is often preferable for a forensic engineer to contract directly with an attorney to protect the confidentiality of the engineer's work. Where the engineer is contracted with a party, the contract should be clear about the accessibility of the engineer's files and the engineer's duty to the general public.
- What would the caller like to have to further evaluate the engineer's credentials?

The initial contact should also determine the next step. Should the engineer send a resume, return a confirming letter, be available for an interview, become familiar with some of the case circumstances before further negotiations, propose terms of a contract, or schedule a meeting to obtain more details of the case and discuss qualifications, scope, schedule, and any conflicts? The State of California requires professional engineers to provide their services with a written agreement spelling out a few minimal issues (California 2012). This includes engineers providing forensic services. During the initial communication, the engineer's ethical patterns, whether good or bad, are likely demonstrated here and normally carry through the rest of the engagement.

One problem in taking an initial call stems from the practice, among some attorneys, of telling prospective consultants things about a case that are privileged or are not already a part of the public record. Once these have been revealed, the consultants may be disqualified from participating in the investigation even though the caller has no intention of retaining their services. This technique may allow an attorney or other client the opportunity to exclude competent consultants, thereby preventing other interested parties from retaining them. In a situation where you receive a call regarding a failure, it may be extremely important to state right away that you should not be told anything that is not public information, and that regardless of what you are told, you reserve the right to work on the case for any interested party.

5.3.3. Conflicts of Interest

Before taking any commission, it is important to rule out the possibility of a conflict of interest. If a forensic engineer works for a large firm, a colleague employee may already have been approached concerning the project. Further, any past relationships with any of the interested parties should be explored and revealed. It is typically essential to ask first for the names of all interested parties so you can determine whether there may be a potential conflict, either real or perceived.

Conflicts of interest fall into two categories, actual conflicts and the appearance of conflict. An actual conflict occurs if, related to the current case, there have been prior involvements of forensic engineers (or anyone in the same firm) with any of the other parties on a professional, personal, or institutional basis. An example of an *appearance* of conflict occurs when the engineer has had previous involvement with one of the parties on unrelated projects or has rendered apparently dissimilar positions in prior similar cases or in publications. In either case, the specific facts of the prior and current involvements determine whether the conflict is real. Apparent conflicts can be dealt with fairly easily and are not a basis for discontinuing the engagement. Bias may be claimed due to prior involvements, and may exist, but does not make an engineer unethical. Failure to set the bias aside in favor of objective evaluation of the case at hand is unethical.

Actual conflicts and insoluble appearances of conflict of interest occur when there exists any influence, loyalty, interest, or other concern capable of compromising the forensic engineer's ability to provide an objective and unbiased professional opinion. ASCE (2000) defines a conflict of interest as "any activity, transaction, relationship, service, or consideration that is, or appears to be, contrary to the individual's professional responsibilities or in which the interests of the individual have the potential to be placed above [the individual's] professional responsibilities." ASCE (2000) further states that members are expected to avoid any relationship, influence, or activity that might be perceived to or actually impair their ability to make objective and fair decisions.

Three degrees of conflict of interest for expert witnesses are implicitly defined in the American Bar Association (ABA) Code of Professional Responsibility (Johnson 1991). The ABA's first degree, or *actual conflict of interest*, pertains to interests that are certain to adversely affect the opinion of the forensic engineer. The ABA's second degree is *latent conflict of interest*. These are interests that may have a reasonable chance of affecting opinions of the engineer. The ABA's third degree of conflict of interest is termed *potential conflict of interest.* These are involvements that can be foreseen to cause a conflict of interest.

One special form of conflict of interest is *bias*, which can be defined as an inability or unwillingness to consider alternative approaches or interpretations (Macrina 1995). Bias cannot be eliminated. It is an unavoidable human characteristic. The forensic engineer has to recognize bias (in himself and others), evaluate the threat which that bias poses to a particular position, and work to overcome, accommodate, neutralize, or otherwise deal with that bias.

Before starting a forensic investigation, the investigation team should provide a full disclosure of potential conflicts of interest to the client's legal counsel. The attorneys can then decide if the potential conflicts are real or are significant enough to discredit the investigative team's objectivity. If they so decide, a different investigative team should be retained to perform the investigation.

Engineers provide services for remuneration. The last type of conflict arises where this interest is subverted, that is, when an engineer cannot serve the public's interest and the engineer's own self-interest simultaneously. It is not valid to call a pecuniary interest a conflict, even though many cross-examining attorneys attempt to do so, because all experts need to be paid for their services. A pecuniary conflict exists when an engineer's fee is contingent on a particular outcome of the investigation. For example, should the client offer to double the engineer's compensation if the client wins the dispute, or should the engineer offer to forego the fee if the client loses, the fee is outcome dependent, and accepting it is unethical.

5.3.4. Confidentiality

Be prepared to keep confidential information confidential. However, because serving the interests of public safety is paramount, even this simple tenet may become a problem. If there is a matter of public safety, the engineer should allow the client a reasonable opportunity to inform the appropriate parties. Failing that, it may be necessary to divulge information that compromises a client's position but which is essential for public safety and welfare. This can be especially challenging, but adherence to this duty remains essential. While it may happen that an engineer is called to task by a client or counsel for revealing something that was learned in confidence, so long as the engineer acted out of a genuine conviction that those actions were serving the interests of public safety, the engineer's actions are defensible.

Although unusual, a particularly challenging case might arise where an engineer is retained by a government agency. If, in the course of his work, the engineer must divulge critical information essential for public safety, what if this information is potentially adverse to the client? Or worse, what if the client is the very agency to whom the damaging information must be reported?

Clearly, the engineer must insure the preservation of public safety and pursue whatever avenues are required to meet the obligation to public safety, possibly including reporting to a different government agency to insure appropriate attention and action.

If information of a confidential nature is provided to an expert by his or her client, the expert should keep records of the consultation and honor the confidentiality. An expert should also notify the retaining lawyer if third parties attempt to discuss a case with the expert.

Communications between the engineer and the client's attorney are normally considered to be confidential and privileged and are usually not subject to being produced during discovery. This rule applies to correspondence or exchanges of written materials that are a part of the overall development of a position, and possibly of a strategy for handling the case, by the client and attorney. Little concern exists here regarding ethical practices, as long as the engineer adheres to that engineer's common practice regarding retention of case materials. Wise choices in what materials are retained in paper or computer files usually obviates any difficult decisions regarding what materials are to be produced in response to an interrogatory or other discovery mechanism. With the exception of privileged materials, the ethical engineer has a duty to produce all the contents of the project files if appropriately asked for these items.

5.3.5. Scope of Services

Every contract should include a scope of the basic services and provision for additional services the client requests. The State of California requires that professional engineers provide their services under the terms of a written contract that must contain five specific clauses (California 2012):

1. A description of the services to be provided to the client by the professional engineer.
2. A description of any basis of compensation applicable to the contract and the method of payment agreed upon by the parties.
3. The name, address, and license or certificate number of the professional engineer, and the name and address of the client.
4. A description of the procedure that the professional engineer and the client will use to accommodate additional services.
5. A description of the procedure to be used by either party to terminate the contract.

In developing a scope of services, it is rarely possible to anticipate all of what is to come. As noted above, it is often appropriate to outline the expected scope and to describe additional efforts that may become necessary later on. One approach is to develop the scope of work in phases. To a degree, it is important that the investigation be complete and comprehensive. However, there may be instances where a preliminary investigation reveals that the client's actions were the most direct cause of a failure. Based upon the advice of counsel, the client may choose to terminate the engineer's services and negotiate a settlement without completing a thorough investigation. However, if the engineer's work is to be complete and comprehensive, it should be based upon an investigation that is sufficient to support its conclusions.

In developing the scope of services, one should avoid language that implies that the work will not be thorough or that it appears to focus only on the client's point of view. Time constraints, fee constraints, and availability of the engineer may limit the final scope, but sufficient resources in all these categories should be incorporated to assure an adequate analysis of the failure and its causes. The scope of work is discoverable, and should be free of language that suggests that the work be crafted to produce a particular conclusion. Feder (1991) recommends that the following elements be included in a typical scope of services for a forensic analysis of a failed facility:

- Planning meeting,
- Understanding a claim,
- Observation, on-site and off-site,
- Document collection, review, and indexing,
- Project analysis,
- Analysis of facility construction schedule (if the failure involved a constructed facility),
- Review contractor's bid and claim documents,
- Participation in legal discovery, proceedings, depositions, alternative dispute sessions, arbitration, or trials.
- Meetings and reports,

Though not included in Feder's list, the engineer should also analyze design drawings, specifications, shop drawings, inspection reports, and records of job meetings.

In coming to a conclusion, the engineer's opinions and conclusions should be able to be supported by rational argument based on evidence obtained through reliable, valid means. Therefore, the scope of the engineer's investigation should be broad enough to assure that adequate time and effort is brought to bear. An opinion that is based on insufficient information is easily undermined and does not serve the client's interest or the court's needs.

The forensic engineer cannot know what information may have been withheld, or simply overlooked. It is especially important, therefore, to be complete and precise about which documents were reviewed, and to state that the resulting opinion is based on information provided. If new information comes to light, the forensic engineer should reserve the right to update that opinion.

5.3.6. Contracting for Services

Elements of your contract have ethical implications. Just as limited-services contracts in design work can result in overlooking important factors that often lead to failure (TCFE 1992), composing an agreement for an investigation that will attempt to prove the client's position can result in overlooking other causes of the failure. Other pitfalls in contracting include:

- Associating compensation provisions with the outcome of the investigation.
- Lack of clarity regarding compensation amounts for appearances (depositions, testimony) versus investigation costs.
- Including a scope of work that would allow or has appearance of advocacy without appropriate, adequate, and thorough investigation.
- Underestimating the level of effort needed to conduct an appropriate, thorough investigation.
- Including unrealistic schedules for completing a proper investigation.
- Accepting a specific, limited scope as the only likely effort that will be contracted.

Numerous textbooks on expert witness work, such as Feder (1991), contain sample engagement letters, sample scopes of services, and sample contract forms.

5.3.7. Compensation

The compensation an expert receives cannot be linked to the outcome of the investigation or the outcome of the trial. Attorneys often accept a case on a contingency fee basis because it is a way representation can be provided to people who have little economic ability to fund it otherwise. This is inappropriate for engineers, because the forensic engineer's role and duty are different from the attorney's. Contingency fees for engineering investigations are prohibited by codes of practice of professional associations and by state licensing boards.

Practically every engineer is asked on cross-examination, "Are you being paid for your testimony?" Experts must be cautious in answering to avoid an appearance of bias or a conflict of interest. The Associated Soil and Foundation Engineers ASFE (1987) recommended the answer, "I am being paid for my time, not for my testimony." Most accomplished experts find that an honest, unrehearsed response is adequate.

5.4. CONDUCTING THE INVESTIGATION

Chapter 4 provides an outline of what a typical forensic investigation might include. This section elaborates on where there may be ethical considerations, pitfalls, or problems. Among the principal ethical considerations in conducting an investigation is to do so objectively and competently.

Section 5.4.1. returns to the story of engineers Q and R with some examples applicable to field work. Ethical behavior is required in all aspects of an investigation in addition to those situations described in the story. For example, it is unethical to:

- Remove evidence from the scene without careful identification and subsequent provenance.
- Alter evidence at the scene.
- Delete or redact non-supportive notes or information from documents, publications, or exhibits.
- Delete portions of data from records.
- Alter or discard photographs or videotapes.
- Withhold relevant material during discovery.
- Be vague about work being conducted or in stating opinions.

This last approbation applies principally to later testimony. While on site, remaining quiet about what one is looking at or into is not unethical. In fact, discussing observations or revealing preliminary opinions extemporaneously could represent a breach of confidentiality.

Failure to conduct a thorough investigation of all reasonably potential causes and effects may result in erroneous conclusions, or at a minimum, may compromise the engineer's credibility when the opposing attorney raises other reasonable causes while the witness is on the stand. Without a thorough investigation, the expert may make an improper linkage of cause and effect or may overlook other possible causes.

A good example of an inadequate cause-effect analysis occurred in the 1998 lawsuit by Texas cattlemen against Oprah Winfrey (National Agricultural Law Center 1998). Their claim was that her televised show on "mad cow" disease caused cattle prices to drop, after the show aired, from $61.90 to $55.00 per hundred-weight. The witness for the cattlemen, a professor from a leading university, stated his opinion that the show's impact depressed cattle prices for at least 11 weeks. Under cross-examination it was learned, and had to be conceded by the witness, that cattle prices usually drop in the same season that followed the show. Winfrey's defense lawyers claimed that the prices could have easily dropped because of season, drought, poor exports, oversupply, or other negative media attention besides their client's show. Of ethical relevance here is whether the cattlemen's expert witness investigated the other possible causes in reaching his conclusion.

Even though the attorneys' objectives of proving their viewpoints and winning the case are important, it is paramount for the engineer to consider as many reasons for the failure as possible. The engineer should not limit the investigation to only those avenues of study that are most beneficial to the client's position, as failures are often due to more than one reason, and often one failure mechanism could trigger another. The forensic engineer should present the truthful technical evidence, developed using valid and reliable means and methods, in the best possible light for the client. There is an important line, however, between casting the engineer's opinion in the best possible light for the client's case and presenting untruthful testimony. It is ultimately up to the jury to decide the case on its merits, but it is unethical for the testifying engineer to be untruthful, intentionally misleading, or dishonest in presenting his opinion.

The scope of the forensic expert's planned investigation should call for evaluation of all reasonable causes. An attorney or client may desire a scope that is limited to investigating a single cause, which may be appropriate but should be debated in the presence of the attorney early in the engagement. If there is a scope conflict among the attorney, client, or the expert, it should be resolved early, or the expert should not take the assignment.

5.4.1. Field Work

If a failure resulted in injury or death, efforts at search, rescue, and recovery have likely altered the failure site. Much of the physical evidence may already have been moved or altered. It may be important to locate elements of construction after a failure, but it may not be possible to confirm that none has been moved or altered. Where there is uncertainty, the engineer's field notes should be explicit in documenting that uncertainty. Wherever there is evidence of disruption to the physical evidence, the engineer should note that some things appear to have been moved.

Returning to the story about engineers Q and R: They bump into each other on site. R is working for the building designer. Both need to measure and estimate the likely snow load on the roof of the damaged building. The building owners tell both engineers that they may not go on the roof of the collapsed building for fear of their safety and out of concern for the owners' liability. Further, the roof geometry has been altered by the collapse, and the roof surfaces are generally no longer at their original angle. Q climbs to the roof of two nearby buildings, makes careful measurements of depth (approximately two feet), removes undisturbed samples of snow from three rooftop locations, and bags the snow for later weighing (16 to 18 lbs. per sample). R ignores the owner's directive, climbs onto the collapsed building, and finds a place where snow fractured along a surface that measures 3 ft. He poses for a photo that shows his tape measure documenting this depth. R then walks to the adjacent field, measures a snow depth of 18 in., and extracts a sample that weighs 30 lbs. R will later argue that in his professional opinion, the roof failed because it was overloaded. Based upon his field investigation, the building was likely loaded to 60 psf—above its design load of 50 psf. Conversely, Q will argue that the building was likely supporting a snow load no greater than 20 psf at the time of its collapse.

If this were a dispute at trial, would it be ethical for Q to explain that he and R were directed to stay off the building but R defied that order? Answer: Yes. Should Q reveal that the fracture surface R measured was clearly a plane that had been at an acute angle to the roof when the building was upright, but which appeared nearly vertical because the collapsed roof sloped steeply beneath it? Answer: Yes. Would it be ethical for Q to add that the field next to the building is low lying, the temperature on the day of the site visit was well above freezing, and the snow was soaked with water where R took his "representative sample?" Answer: Yes. In light of these additional facts, is there anything about R's behavior that falls within the bounds of ethical practice? Or does it reflect his acknowledged bias that he would never take on an assignment that might place him in opposition to designers and contractors who specialize in this building type?

In evaluating the ethics of R's behavior, one has to separate the question of R's use of invalid and unreliable methods to support his opinion, which in themselves could be reasons to discredit or exclude R's testimony (or challenge his competence) from the question of R's intentional manipulation of sampling techniques to favor the side that hired him. Such manipulation may very well be a breach of ethics.

Although R may have acted unethically, he would likely argue that he appropriately determined the depth of snow by measuring it directly on the collapsed roof rather than relying on measurements on nearby buildings. He determined the density of the snow by sampling it nearby. Then, by combining the depth and density, he presented a reasonable engineering estimate of the likely snow load. He might further argue that he chose to measure depth in a place that seemed deep because he was concerned that he not underestimate the snow depth. Further, he might argue that his sample in a nearby field represented his best effort to determine the density of the snow. In both cases, he was acting in his client's interest by making sure he found the deepest place to measure and the heaviest density. He was hired to help his client, their interests are clearly well served by a conclusion that their building was simply overloaded and was not deficient in any way. How, then, can any of what he did be considered unethical? Perhaps Q sought out those places on the nearby buildings where the snow depth was least. Further, by extracting and weighing the samples as he acquired them, perhaps Q was missing the weight of any moisture that may have melted at the base of the snow pack and run toward a drain. It is the jury's job to weigh the evidence, including the validity and reliability of it. The jury must weigh the credibility of the witnesses and their testimony and answer the ultimate question. It is proper for experts to testify regarding the validity or reliability of another expert's testimony.

5.4.2. Laboratory Testing

Whenever laboratory testing is planned, if the investigation is part of a known dispute, it is common for all interested parties to be present during testing. This is almost always essential for destructive testing, but likely applies whenever physical evidence will be manipulated and may be altered (even, for example, by cleaning to prepare a specimen for microscopic examination). So long as the principal investigators possess the appropriate expertise, they may undertake the testing. However, the laboratory is a common place to use the services of someone capable of serving as an expert in their area of practice. A structural engineer may well understand basic metallurgy, for example, but a metallurgist or a certified welding inspector could be a more appropriate choice to conduct a microscopic examination of a broken weld.

5.4.3. Analyses

Office work may range from simple hand calculations to sophisticated computer analyses. Even simple calculations, based upon only a few assumptions, may lead to significant differences of opinion. Computer models typically rely upon myriad assumptions and may lead to complex arguments about the underlying assumptions and methods of approximate analysis. All analyses should be undertaken with objectivity. Where a particular analysis is sensitive to the underlying assumptions, a parametric test of those assumptions is appropriate. It is rarely possible to conduct all analyses for any particular investigation, but those chosen and conducted should be defensible and credibly complete

A structural engineer once commented that if he thought a finite element analysis might show that his client was at fault, he would not undertake that analysis. Although this may appear unethical, if the engineer explained to the attorney that such an analysis would likely lead to an adverse conclusion, and if the attorney used that information to settle a case early, the act of omission might well be ethical. If the analysis was omitted without explanation, and later, because it had not been done, the engineer argued that there was nothing in his analyses that implied fault with his client's work, the omission would be unethical. In this case, the engineer would be intentionally misleading the court. However, if, by undertaking such an analysis, he found that its results were heavily dependent upon simplifying assumptions, and that those assumptions are not necessarily appropriate in his case, then he could and should undertake the analysis, test for its sensitivity to those assumptions, and present his results objectively. If, in his professional opinion, there is enough latitude in the way the assumptions can be applied, he may be thorough and objective, while simultaneously helping support his client's position.

In the story of Q and R, Q went through the precipitation records for two nearby communities. Starting from the last date when temperatures were above freezing, and adding all precipitation recorded from that date until the date of the collapse, he found that there had been an equivalent of 8.4 in. of water—although this had fallen both as snow and as a mixture of snow and rain. If all of that precipitation had landed on the roof of the subject building and if none had either melted or been blown off, the resulting load on the roof would have been less than 44 psf. R ignored this data. Because there were no records kept at the failure location itself, he argued that the vagaries of local weather patterns and precipitation amounts confirm that we do not know how much snow was on the roof at the time of the collapse. Therefore, the best estimate of snow load on the roof must come from direct measurement—his.

5.5. TESTIMONY

Expert witnesses may well perform investigations, report their results and conclusions, and learn that their attorney client has settled the dispute without ever requiring expert testimony before a judge and jury. However, experts must always anticipate that their work may lead to a requirement for testimony under oath, whether in a reference hearing (common in insurance dispute resolution), in mediation (where testimony may or may not be given under oath), in arbitration, or finally, in a courtroom before a judge and jury. The ethical issues present in testimony are many and varied.

This chapter has argued that a forensic technical expert must be objective and impartial. Further, one may test oneself against this standard by asking whether one's opinion would be different if the client was on the opposing side. If all experts are objective and impartial, and each is forming an opinion based upon the same set of facts, with the same analytical tools at hand, how then, can two opposing experts in a dispute reach different and opposing opinions in the case? A paper by Kardon, Schroeder, and Ferarri (2003) addresses this question, and provides a lengthy treatment of the subject. They conclude as follows:

> *The expert's obligation extends beyond the expert's client, and includes a duty to opposing parties and to the larger society as a whole. As a consultant, prior to being disclosed as an expert, the technical forensic specialist must thoroughly investigate the dispute from all sides, and inform the retaining attorney of the strengths and weaknesses of the arguments from all sides, given the consultant's specialty and assignment. When then disclosed and called to testify, the expert answers all questions truthfully, without going beyond the scope of the assignment and without volunteering information or opinions not called for by the questions posed.*
>
> *What serves justice best is competent expert testimony explaining technical aspects of the dispute from the point of view of each party. The expert is not the trier of fact, but instead must "assist the trier of fact to understand the evidence or to determine a fact in issue."*

In that role, the expert for each side must evaluate the available evidence and present those legitimate opinions and facts which reasonably support that side.

Among other clear ethical dilemmas, the testifying expert is sworn to "tell the truth, the whole truth, and nothing but the truth." It is unlikely that anyone will know "the whole truth," regardless of their expertise and the thoroughness of their investigation. However, it is absolutely necessary that the ethical expert tell all the truth that the expert knows and that is asked about, tells only truthful things, and tells nothing but the truth. Within the confines of the courtroom (or any other dispute-resolution setting), the expert is also often directed to answer the question as it was asked or to only answer the question posed. Therefore, the expert's answers will depend upon the skill of the questioning attorney. It is not appropriate to take advantage of a poorly worded question to mislead the trier of fact. However, it is appropriate to ask the questioner to pose the question in smaller bits, if it is unclear, ambiguous, or poorly framed. Returning to Kardon, et al.:

In the adversarial system of litigation, it is principally the attorney's responsibility to argue the case, not the expert's; it is the jury's responsibility to determine questions of fact, not the expert's. It is not the responsibility nor the prerogative of an expert on one side of a case to develop and volunteer testimony in favor of all sides.

5.6. ASCE CODE OF ETHICS

For background on ASCE's codes of ethics, the reader is referred to a comprehensive review by the society's General Counsel (Smith 1999). The American Society of Civil Engineers adopted its original code of ethics in 1914 and has updated it in 1927, 1961, 1964, 1975, 1977, 1980, 1996, and 2009 (ASCE 2012, ASCE 2000, Smith 1999). In the Code of Ethics, the only specific mention of conduct while serving as expert witnesses is in Canon 3.c. The complete Code is included as the Appendix to these *Guidelines*.

Canon 3.c of the ASCE code relates to three ethics categories, thoroughness of investigation, competence, and honesty. It states:

Engineers, when serving as expert witnesses, shall express an engineering opinion only when it is founded upon adequate knowledge of the facts, upon a background of technical competence, and upon honest conviction.

The ASCE Code of Ethics has been reviewed for relevance to current practice (Groden 1991, Smith 1999) and by the TCFE Ethics Committee (TCFE 1992). TCFE examined the Code relative to forensic engineering and assessed its applicability to the future. The committee determined that the Code *was not only adequate but also exemplary in its definition and objective.* They concluded that Canon 3.c was the principal operative for forensic engineers, and that unethical practice is usually traced to the lack of adequate knowledge of the facts or standards, lack of technical competence, or lack of honest conviction (TCFE 1992).

It is interesting to note that TCFE's Ethics Committee also concluded, in the context of competence, that testimony by university professors and research scientists who do not have a background in forensic practice should be restricted to their narrow area of expertise and should exclude statements regarding the cause of failure. This is not a statement of ASCE policy, but communicates the concern by many design practitioners that a prerequisite for testimony regarding failure of designed facilities is experience in design. Those associated with a university or research center considering engagements as experts should first review the qualifications to practice engineering in their states to determine if expert engineering testimony is considered "engineering" which would require licensing or registration as an engineer. An engineer should also consider the ramifications of testifying regarding engineering design if the engineer's background has not included engineering design, or exposure to design standards.

5.7. DAMAGE DONE BY UNETHICAL CONDUCT

Numerous instances in the forensic expert system could be cited to illustrate the damage being done to the engineering profession (and to the public) by careless or sloppy work by incompetent or unethical engineers and scientists. As an example, an incident reported in the press (*USA Today* April 16, 1997) makes this point. The article describes the findings of the Justice Department report on "shoddy work, lax procedures and questionable practices" by experts in the U.S. FBI Laboratory (USOIG 1997). According to *USA Today*, the Justice Department findings in regard to investigations and testimony by the "experts" included:

1. "Scientifically flawed testimony" in four cases, including the bombing of the World Trade Center garage.
2. "Inaccurate testimony" by lab examiners in three cases, including the World Trade Center garage case.
3. "Testimony beyond the examiner's expertise" in the World Trade Center garage case and in the investigation of the 1989 crash of Avianca Airlines Flight 203.
4. "Improper preparation of laboratory reports" by technicians in the explosives unit.
5. "Scientifically flawed reports" in the investigation of the Oklahoma City bombing and other cases.
6. "Inadequate record management and retention system" by the lab.

Shoddy work, lax procedures and questionable practices can result in reduced credibility of the expert witness system, and increased challenges to the veracity an importance of forensic engineering practice. In 1996 alone, the lab analyzed 600,000 pieces of evidence in 20,000 cases. Numerous challenges of these and thousands of other technical cases and convictions could be the result, causing harm to the reputation of the engineering profession, and incurring expenses on the part of taxpayers and of the parties in the cases.

5.8. REPORTING UNETHICAL CONDUCT

When anyone observes conduct or services of forensic engineers that appear to disregard the safety, health, and welfare of the public, that observation should be reported immediately. This is a duty, not an option, of all engineers. Serious or repeated breaches of any of the principles or canons of the ASCE Code of Ethics should also be reported. Violations of the ASCE Code of Ethics should be reported to the ASCE's Committee on Professional Conduct (CPC). Violations of state licensing laws should be reported to the state board or agency responsible for registering or licensing engineers.

The incidence and composition of engineer's reports of possible unethical conduct is unknown. Each state registration or licensing board normally publishes a list of disciplinary actions, but the Committee is unaware of the availability of a compilation of national statistics of such actions. It is likely that the percentage reported is small.

Once unethical conduct or services have been reported, it is the duty of the licensing boards, or other authority to examine the particulars of the case, allowing these to be presented, heard, and judged. If warranted, the governing authority may then take action against the practitioner.

Most states require engineers to report incidences of what may be perceived to be unethical behavior. Aside from state requirements, this duty also arises from membership in a professional society, such as ASCE, which requires the self-policing of its members through its Code of Ethics. To fail to report does not serve the public's interest and may in itself be unethical

In conclusion, adherence to the principles and ideals discussed in this chapter by forensic engineers should improve objectivity and mitigate the damage being done to the public trust and to the profession by engineers who engage in unethical practices. These *Guidelines* encourage discussion in ASCE's journals on the ethical issues and compiled principles listed in this chapter, and encourage suggestions for improvements in professional engineering support of litigation.

5.9. ETHICAL STANDARDS FOR PUBLICATIONS

The ASCE has promulgated ethical standards in its *Authors' Guide to Journals, Books, and Reference Publications* (1997), including lists of ethical obligations of authors, reviewers and editors (ASCE 1997). While many of the canons of ethics presented for authors refer to avoidance of fragmentation and submittals of the same material to multiple journals, others relate to objectivity, competence, plagiarism, falsification of data, and the difference between personal and scholarly criticism. The position of the FPC as presented in these Guidelines is that the members of ASCE who prepare written documents for use in court or administrative proceedings should be familiar with and comply with the ASCE Code of Ethics, and comply with ASCE's *Authors' Guide to Journals, Books, and Reference Publications.*

5.10. CONCLUSION

When serving as an expert witness, the forensic engineer is often at the center of our adversarial legal system. The very definition of *forensic* derives from the notion of discussion or debate in a public or legal forum. However, as defined in our court system, an expert witness is seated at trial to aid the trier of fact, to present technical points in a way that is accessible to a lay jury, and to express opinions based upon sound engineering principles and objective engineering opinions. When a forensic engineer drifts away from the duty of objectivity towards advocating on behalf of the clients' cause without adequate investigation and analysis using valid and reliable methods, or when the forensic engineer shades the truth or misleads the jury in an attempt to win a case despite the absence of valid, supportable evidence, this represents an ethical breach.

5.11. REFERENCES

ASCE, American Society of Civil Engineers, *ETHICS: Standards of Professional Conduct for Civil Engineers,* New York, April 3, 2000.

ASCE, American Society of Civil Engineers. *Code of Ethics,* Reston, VA, July 23, 2006. (See Also http://www.asce.org/Content.aspx?id=2147484014)

ASCE, American Society of Civil Engineers Web Site: http://www.asce.org/ (May 24, 2012)

ASFE, Associated Soils and Foundation Engineers,. *EXPERT: A Guide to Forensic Engineering and Service as an Expert Witness,* Assoc. of Soil and Foundation Engineers, Silver Springs, MD, 1987.

California Business and Professions Code §6700-6799, Professional Engineers Act, Sacramento, CA, 2012.

Commonwealth of Massachusetts, Regulatory Authority 250 CMR 4.00: Massachusetts General Laws c. 112, §§ 81D through 81T, (2012).

Feder, H. A. (1991). *Succeeding as an Expert Witness*, Van Nostrand Reinhold, New York.

Friedlander, M.C. (1990), *A Radical Proposal to Stem the Professional-Liability Crisis*, Consulting/Specifying Engineer, June.

Groden, B.T. (1991). Is the ASCE Code of Ethics Obsolete in Today's Society? *Civil Engineering Magazine*, January.

Johnson, D. (1991). *Ethical Issues in Engineering*, Prentice Hall, New Jersey, p.392.

Kardon, J.B., Schroeder, R.A., Ferrari, A.J. (2003). *Ethical Dilemmas of Technical Forensic Practice.* Proceedings of the 3rd Forensic Congress,, Technical Council on Forensic Engineering, American Society of Civil Engineers, October 19-22, 2003. San Diego, California.

Macrina, F. (1995). Scientific Integrity Appropriate? : An Introductory Text with Cases, ASM Press, USA, 283 p.

Smith, T.W. III. (1999). ASCE General Counsel, *ASCE ETHICS, Edict, Enforcement, and Education*, American Society of Civil Engineers, 1999 Zone Management and Leadership Conference.

TCFE. (1992). Minutes – Technical Council on Forensic Engineering, New York, Sept. 12.

United States Office of the Inspector General, Department of Justice. (USOIG) (1997, April). The FBI Laboratory: An Investigation into Laboratory Practices and Alleged Misconduct in Explosives-Related and Other Cases.

Chapter 6
The Legal Forum

6.1. INTRODUCTION

The American system of jurisprudence provides for resolution of disputes by a structured, but seemingly arcane process. The system presumes each side of a dispute may be presented by professionals trained in the process (lawyers), to a panel of one's neighbors (the jury), overseen by another professional (the trial judge), who orchestrates the process (the trial). This all occurs within a set of rules created either by the legislature or by a body of commentary created by appellate judges, who do not actually oversee trials, but instead review and evaluate the trial judges' conduct to provide a guide for other trial judges. Having heard both sides, the jury resolves the dispute. This is litigation. Civil wrongs, known as "torts," arising out of conduct deemed negligent are adjudicated within the court system by litigating whether the conduct was negligent, and whether it proximately caused the injury. Juries hearing disputes involving negligent misconduct often and assisted by opinions offered by forensic engineers.

This chapter briefly describes the American system of jurisprudence and discusses the relationships of the forensic engineer with other parties in the system. After these introductory sections, the chapter focuses on admissibility of testimony from a legal perspective and discusses the role of engineers in non-adjudication forums.

This chapter is not intended to serve as a thorough exposition of these topics, but rather to introduce some of the key terminology and outline some of the legal system workings related to forensic engineering. Readers seeking more thorough descriptions are referred elsewhere.

6.1.1. Description of the Legal System

Litigation is generally defined as a "lawsuit." It begins with the official filing of the complaint within the appropriate court system or alternate dispute resolution forum. As the lawsuit proceeds, each party is entitled to engage in a process called "discovery," where all evidence and testimony any party may offer at trial can be fully examined by all parties.

The discovery process is intended to avoid a trial by surprise. By reviewing the total body of evidence, the parties and their attorneys are able to gauge the potential for courtroom success. Perhaps if each party is fully informed of the facts and legal bases available to the other side, and is fully aware of the facts and legal bases available to it, and can gauge the potential for success at trial, reasonable people will reach a common ground prior to trial, and the case will be resolved without trial (commonly called a "settlement"). Much of the evidence gathered during the preparation of a case based on a tort claiming negligence, or a case with an engineering context alleging a product defect or error in structural design, is gathered by opinion witnesses or reviewed by opinion witnesses, and many of those opinion witnesses are forensic engineers.

In many jurisdictions, construction cases can be designated as complex cases—those involving many complicated issues requiring extensive discovery. For instance, interrogatories—a discovery tool used early in the litigation to discover the names of witnesses and location of pertinent documents—are written questions answered under oath by a litigant and are passed between parties. A defendant may request that an injured plaintiff name all occurrence or post-occurrence witnesses, list all treating or consulting physicians and all hospitals or clinic where treatment was rendered, and perhaps outline the proof of the plaintiff's case, particularly disclosing any expert opinion witnesses the plaintiff may call. A plaintiff may request that the defendant list all persons involved in the design, fabrication, or construction of the product or structure, all applicable Standards or Codes, any prior complaints, the defendant's opinion witnesses (both those who may have been retained as consultants to the defendant and those who are employees of the defendant), and outline the theory underlying the defendant's assertion of non-responsibility.

The discovery process in a complicated construction case may last for a year or more. Witnesses, both fact and expert, are identified in the interrogatory answers, after which witnesses are routinely examined in a proceeding known as a deposition. In the deposition, attorneys for the opposing parties subject the witness to extensive questioning in a fairly informal setting, perhaps the lawyer's office. The witness is under oath, and a court reporter makes a written and perhaps a video record of the proceeding.

A third tool of the discovery process is a "request for production of documents," written requests passed between parties, seeking copies of pertinent documents in the possession of one party or the other, or in the alternative, the locations of such documents which then may be acquired by a court-mandated process called "subpoena." The individual discovery tools are discussed below.

Once discovery is completed, the case may be settled or it may proceed to trial. Because of burgeoning caseloads, many jurisdictions require litigants to attempt to resolve the dispute by a mandatory settlement process known as mediation. In this forum, the parties hire an outside negotiator to aid in reaching a settlement. This process is "off the record," and no information discovered may be used at trial. This process ends with a mutually agreed upon settlement or when it becomes clear that a negotiated settlement is not possible. It is not unusual for the mediator to stay involved with the case right up to and sometimes during the trial.

If the conflict cannot be resolved by mediated settlement, and if the parties choose not to use other alternate resolutions techniques, the case will be tried in court before a judge and jury, or only a judge, sitting as both judge and "finder of fact." Each side presents its demonstrative (non witness) evidence. Witnesses testify under oath. The initial testimony by a witness is referred to as direct testimony. Lawyers for opposing parties may then cross-examine the witness. Further questioning may occur and is referred to as redirect (which is followed by re-cross examination, and so on, until the witness is excused). The plaintiff—the party bringing the suit—has the burden of going forward, presents its suit first. When it has presented all its witnesses and has offered all its demonstrative evidence to the court, it "rests." The defense then presents its case-in-chief.

Arbitration and mediation are non-judicial forms of litigation. Instead of the judge and jury, one or more arbitrators or mediators hear the case. A panel of arbitrators is usually called a tribunal. The arbitrator or mediator may be a lawyer, engineer, architect, or construction expert. Usually a tribunal will contain a mix of professional backgrounds.

Arbitration hearings and mediation proceedings are less formal than those of the courtroom. Stenographic records may or may not be taken. In arbitration, the lawyers present their cases just as they do in the courtroom. The arbitrator may also pose questions to the witnesses. Opinion witnesses often do not appear in person, but instead submit their work product to the arbitrator or mediator in the form of an affidavit or report, or, the opinion witness may have appeared earlier at the lawyer's office and questioned as if at trial, in a process called an "evidence deposition" (described more fully in a later section). The written transcript of the evidence deposition will be submitted to the tribunal as the opinion witness' testimony.

Arbitration or mediation is desirable for a number of reasons and used more and more often to resolve disputes. The cost to the parties is significantly lower than formal litigation. The discovery process can be streamlined and expedited, saving attorney fees. The time it takes to arrive at an award is reduced. In some tribunals, construction industry personnel hearing the case are more likely to understand the facts than an ordinary jury. The decision to arbitrate or mediate can be a contractual or legal issue, but normally, if not provided for in the contract documents originally binding the parties, the parties must agree to an alternate dispute resolution method.

6.1.2. Forensic Engineers as Experts

The court refers to engineers working within the legal system as expert witnesses or opinion witnesses. An expert is someone who knows considerably more about a certain topic than an ordinary citizen. Federal Rules of Evidence Rule 702 reads as follows:

> *If scientific, technical, or other specialized knowledge will assist the trier of fact to understand the evidence or to determine a fact in issue, a witness qualified as an expert by knowledge, skill, experience, training or education, may testify thereto in the form of an opinion or otherwise, if (1) the testimony is based upon sufficient facts or data, (2) the testimony is the product of reliable principles and methods, and (3) the witness has applied the principles and methods reliably to the facts of the case.*

The State Court criteria determining qualification as an expert witness is similar to the Federal Rule in most States. Using those parameters, the judge or arbitrator decides whether the witness is qualified as an expert. If so, the judge will permit the qualified expert to testify and, most importantly, offer opinion evidence on a wide range or topics, including perhaps the ultimate issue: Was the defendant negligent? Ordinary fact witnesses can only testify based on their opinion in a very limited fashion (for instance, "speed of the other car" assuming they are a licensed driver and had a clear view).

The legal forum is the domain of the lawyer. Forensic engineers are relegated to subordinate roles—that of litigation support offering opinion testimony at trial which may further the lawyer's theory of the case. Forensic engineers are extremely valuable to the legal system; as stated in Federal Rule 702, they "assist the trier of fact to understand the evidence or to determine a fact in issue."

6.2. ROLE OF THE FORENSIC ENGINEER AS A WITNESS IN LITIGATION

6.2.1. Pre-Trial Elements

6.2.1.1. The Report

Federal Rules of Civil Procedure (2011), Rule 26(a)(2) covers "disclosure of expert testimony." Sub-section (B) requires the lawyer to disclose a witness who is retained or specially employed to provide expert testimony in the case and with that disclosure, to provide a written report prepared and signed by the witness. The report must contain a complete statement of all opinions to be expressed by the witness, the basis and reasons for those opinions, and the information considered by the witness in forming the opinions; any exhibits to be used as a summary of or in support for the opinions; the qualifications of the witness including a list of all publications authored by the witness within the preceding ten years; the compensation to be paid for the study and testimony; and a listing of any other cases in which the witness has testified as an expert at trial or by deposition within the preceding four years [FRCP 26(a)(2)(B)].

Many state courts follow the Federal Rule, but some do not. A report is not required in every State. The work product of the forensic engineer generally consists of 1) the investigation, 2) the written report, 3) attendance at deposition, and 4) trial testimony. There may be ancillary tasks a forensic engineer will be called upon to render (assisting in preparing written discovery requests or assisting in preparing the retaining attorney for cross-examining the opposing party's retained expert), but for the most part, the investigation, report, deposition, and trial testimony will be the bulk of the forensic engineer's effort, and of those, the written report will almost always be required.

The report allows the attorney to understand the engineering issues and the opinions of the forensic engineer and helps the attorney plan trial strategy. The report will, to a large extent, contain the facts and opinions to be elicited during direct testimony at trial. But it will also be pored over by the opposing attorneys to plan their cross-examination at deposition and trial. While the engineer who is thinking of entering the field of forensic engineering may be accustomed to presenting proposals to a band of leery-eyed corporate project managers and "bean counters" in an industrial setting, the legal forum can be far more demanding, because it may be intrinsically adversarial and confrontational. Any error or omission is seized upon by the opposing attorney and used to attack the credibility of the forensic engineer.

6.2.1.2. The Request to Produce the Forensic Engineer's "File"

It is common for the opposing attorney to request a copy of the forensic engineer's entire investigative file. But "file" no longer is limited to a folder of paper with printing or writing on it. It now includes digital recorded information, whether stored on a mass media device in the forensic engineer's office or "in the cloud" on a remote server accessible through the internet. Rule 45 of the Federal Rules of Procedure states that a subpoena may be issued commanding each person to whom it is directed to attend and give testimony or to produce and permit inspection copying, testing, or sampling of designated books, documents, and electronically stored information ("e-discovery").

The basis of e-discovery in the Federal System (discovery of electronically stored information) for the forensic engineer is Rule 45 (FRCP 2011), the rule governing subpoenas. Rule 45 says electronically stored information is added as a category of information that may be sought by subpoena [Rule 45(a)(1)(C)]. The subpoena may specify the form or forms in which electronically stored information is to be produced. The default form of production for electronically stored information is "a form or forms in which the person ordinarily maintains it or in a form or forms that are reasonably usable." The court may specify the conditions upon which such discovery shall proceed. A claim of privilege or protection as trial-preparation material may be made after production of discovery material by notice to the receiving party, and the receiving party must then promptly return, sequester, or destroy the specified information and any copies, may not use or disclose the information until the claim is resolved, and, if the information has been disclosed before receiving notice, must take reasonable steps to retrieve the information that was inadvertently disclosed. Generally nothing in the forensic engineer's file is privileged against disclosure.

6.2.1.3. The Deposition

Depositions have at least a three-fold purpose: First, they allow the opposing parties an opportunity to fully explore the forensic engineer's opinions that will be offered at trial. Secondly, the deposition gives parties a chance to determine the potential credibility of the witness. Third, the deposition "freezes" the testimony of the engineer. The deposition testimony can be lengthy and usually involves tremendous detail. Some state rules limit the length to a certain number of hours, but often the judge will permit the attorneys to go beyond these set limitations because of the importance of the forensic engineer's testimony. For the forensic engineer, this process can be laborious and stressful. Even under these very taxing circumstances, the forensic engineer should strive to maintain professional decorum.

Ethically, and because the deposition is an examination under oath, the forensic engineer has but one answer—the truth. Forensic engineers should make sure they understand the question being posed. Questions with multiple parts should be addressed one part at a time. Questions that involve conflicting circumstances should not be answered. The forensic engineer should not answer the question the lawyer meant to ask but should answer what has been asked. After answering the question that was asked, the expert might then have an opportunity to expound further on the topic of the question. The forensic engineer can aid this process by asking the lawyer to rephrase or clarify ambiguous, unclear, compound, or nonsensical questions.

6.2.2. Trial Testimony

While the deposition may be lengthy and can focus on intricate detail, trial testimony generally involves key issues and themes. Ideally, litigation is a search for the truth. The trial is a test of the legal team's ability to make complex issues understandable and interesting. The attorney's biggest challenge is to hold the attention of the judge and jury while laying the factual basis for the positions taken in the case.

It is during the trial that the lawyer orchestrates the witnesses and evidence into a plausible and cogent presentation. The sequence of witnesses coupled with tactical considerations will dictate the order of the presentation, although the plaintiff has the burden of proceeding first, the "burden of going forward." The defendant or respondent participates by cross-examining the plaintiff's witnesses, but not until the plaintiff "rests" his case is the defendant required to go forward with the defense. Often times, the forensic engineer is required to be present in the courtroom waiting to testify or to listen to others (sometimes to assist the lawyer, sometimes to hear testimony of witnesses upon which the forensic engineer's testimony will be based, but often merely because the judge wants the witnesses close at hand so one may quickly follow the other and the trial is not delayed). Such a requirement to be in the courtroom varies from one jurisdiction to another. In some court venues, witnesses may be excluded until they have testified (so they cannot rely on earlier testimony). If the forensic engineer is required or permitted to attend or is allowed to sit with and assist the lawyer during cross-examination of the opposing parties' technical witnesses, the forensic engineer should maintain a professional decorum. The jury will be watching.

While on the witness stand, the forensic engineer should answer truthfully all questions posed. The answers should be given with the level of detail intrinsic to the questions. Because the jury is the finder of fact, the forensic engineer should respond to questions in a way that will be understood by laypersons. If the lawyer properly prepares the case, the forensic engineer witness and lawyer will have discussed the questions that will be asked, and the forensic engineer will have rehearsed the answers. The word "rehearse" is not meant to connote collusion or unethical predetermination and manipulation of testimony. The purpose of the testimony is to assist the trier of fact to understand and resolve an engineering or technical issue of fact. The burden on the lawyer is to fashion the questions so 1) the answers state the expert's opinions and conclusions, and the bases therefore, and, 2) the questions stay within the bounds of the Rules of Evidence. If an engineer entered a production meeting at a plant, the supervisor might say, "Tell us what you did, and tell us what you think and why." The rules of evidence do not allow such informality in court testimony. A question requesting a narrative is not permitted. The forensic engineer's work product is valuable, and it is important that it be presented as fully and as completely as necessary to best inform the jury. The expert witness should insist that the attorney client provide questions (the "direct examination") well in advance of trial.

The bane of the forensic engineer's trial testimony is cross-examination by opposing counsel. Cross-examination at trial is likely to be less onerous than cross-examination during the discovery deposition. The deposition likely exposed the weak points in the forensic engineer's presentation, and the lawyer who hired the forensic engineer will have attempted to lessen their impact by admitting the problems during direct examination in a manner which suggests to the jury the witness is candid and truthful. The opposing lawyer cannot meander about at trial, wandering up and down blind alleys as may have been his strategy during the deposition, but must in most jurisdictions limit cross-examination to the areas covered in direct testimony. So here again, preparation is key. The forensic engineer witness and the attorney client can fashion direct examination in a way to highlight all the engineer's best work, and perhaps leave other less important opinions "on the cutting room floor."

But the forensic engineer as an expert witness should be aware that the attorney who took the expert's deposition may have been the young associate. The attorney facing you in the courtroom may be the crafty, experienced trial partner, who may be quite aware of what is on cutting room floor, because it may have been brought up during deposition or otherwise learned during discovery. It is the duty of the cross-examiner to attempt to undermine the credibility of the forensic engineer. Cross-examination is intended to test a witness's knowledge and recollection or disclose bias or prejudice. Cross-examination of an expert witness is intended to test and probe the witnesses' opinions and the bases for those opinions. Courts have held counsel should be permitted to attempt to discredit or demonstrate weaknesses in the expert's offered opinions and be granted the widest latitude in the process. The Rules of Evidence are intended to allow only the "best evidence" to reach the jury, and there is a school of thought that argues that if the expert's testimony holds up after being tested by the crucible of cross-examination, the jury benefits.

A resourceful opposing attorney may want to "pull out all the stops" in this endeavor. Cross-examination can be extremely stressful. A forensic engineer (or any witness) has no control over what questions may be asked. The witness is obligated, however, only to answer those questions that he or she feels qualified to answer.

Forensic engineers do not win or lose trials; lawyers do. The forensic engineer's role at trial is to present opinions in an understandable and credible way and to assist the trier of fact in resolving technical and scientific issues by providing them with the benefit of the forensic engineer's knowledge That is the ultimate determination of success as an expert witness, not the outcome of the litigation.

6.2.3. Rebuttal Testimony

During trial, the attorney may elect not to call the forensic engineer as a witness on direct examination. This is a legal decision and is driven by strategy. It is not unusual in those circumstances for the forensic engineer to be asked to serve as a rebuttal witness. Usually this involves listening to other expert testimony and then being called as a witness.

A forensic engineer should agree to serve in this capacity only if the engineer has complete knowledge of the engineering aspects of the case. To agree to testify on the spur of the moment with a very limited knowledge of the facts creates a situation conducive to an appearance of unethical practice. The testifying engineer can attempt to avoid that appearance by limiting the testimony to be based only on the available facts and by incorporating the extent of that knowledge of the facts in the answers.

Testimony on rebuttal follows the same format as direct and cross-examination, with one exception. That is, no new facts are allowed to be placed into evidence. The forensic engineer's rebuttal testimony must be based on information previously offered, including testimony by opposing experts.

6.2.4. Arbitration Testimony

A multi-judge or multi-person panel known as a tribunal usually hears construction cases that go to arbitration. Most often, at least one of the arbitrators has technical expertise pertaining to the contested issues.

Arbitration is somewhat more relaxed than a trial in a courtroom with a judge and jury. A stenographic record is optional. The court-established rules of evidence are suspended, and a set of rules is adopted.

Because the tribunal is both judge and jury, the forensic engineer should gear the answers to the technical level that is appropriate. Once the direct and cross-examinations are completed, the tribunal will often engage in further questioning. It is not unusual for the tribunal to decide a case based on the technical merits. The forensic engineer should therefore prepare for testimony accordingly.

6.2.5. Relationship to the Legal Team

It is the function of the legal team to present the best possible case to the court. To do less is to perform a disservice to the client. In the development of strategy, often the lawyers will want to discuss the issues with the forensic engineer. These may include discovery requests, interrogatories, technical issues, and even the approach to cross-examination of opposing experts.

These types of services by the forensic engineer are ethical only if the engineer maintains objectivity. There is nothing wrong with assisting the non-technical attorney in both understanding the complex issues and helping expose the weaknesses of adversarial experts.

It is, however, considered unethical for a forensic engineer to "craft" opinions to benefit the case in the absence of valid, reliable evidence. Conversely, it is acceptable for the forensic engineer to modify the language needed to communicate opinions such that the non-technical public may more easily understand it. There is a distinction that should not be overlooked. The forensic engineer should not alter, expand, contract, or develop technical opinions to aid a client's position in the absence of valid, reliable, truthful technical evidence to support those opinions. The forensic engineer should base opinions on knowledge gained by valid, reliable methods of investigation, testing, evaluation, and analysis. The forensic engineer may ethically substitute more basic or even non-technical language if it assists in understanding.

Although most attorneys will develop their own legal strategy, they lack sufficient knowledge to fully understand the technical engineering or scientific concepts. Consulting with lawyers on strategy is a perfectly acceptable undertaking by a forensic engineer. What experts are needed and helping the attorney to find the most knowledgeable expert, what discovery should be undertaken and helping to design the written and oral discovery to best advantage, how the courtroom case should be structured, or even how to attack an opposing expert's conclusion, opinions, qualifications, or credibility at deposition or trial (In this case, ethically means based on a sound factual and scientific basis, suggesting genuine and germane questions on those elements of the opposing expert's work product as set forth in 6.2.6., below.) are all examples of questions the forensic engineer can ethically address.

6.2.6. Relationship to Adversaries

It is perfectly legitimate for different engineers to draw different conclusions and opinions from the same facts. Different levels of education and professional experience often guide forensic engineers in somewhat different directions. It is unethical for an expert to ignore facts that are contrary to the expert's opinions. The forensic engineer must maintain objectivity throughout the process and is obliged to reevaluate opinions when new facts are discovered.

It should be obvious that the dispute is between the litigating parties—not between the experts. That is, the forensic engineer is retained to aid the trier of fact and is not a party to the dispute. The forensic engineer should establish and maintain a professional demeanor and respectful interface with both the attorneys and experts for the other side.

It is not unusual during depositions or even at trial for an attorney to assail a witness. In depositions, this tactic can be used to test whether the witness will get angry or lose composure. The forensic engineer is well served to handle such an attack with poise and aplomb. Attorneys who attempt this tactic in the courtroom are risking their own credibility with the jury. Although the direct affront can be taxing, the forensic engineer must always refrain from even the appearance of arguing with opposing lawyers. By taking part in the debate, the forensic engineer seemingly becomes part of the dispute rather than fulfilling the obligation to aid the trier of fact.

The forensic engineer's relationship with opposing experts should be that of equal respect. Simple disagreement of opinions is not justification for unprofessional behavior. Opposing experts are no more part of the dispute than the forensic engineer is. Cordiality and decorum are to be expected and are due.

6.3. ROLE OF FORENSIC ENGINEERS IN MEDIATION

Mediation is a non-adjudication method of dispute resolution. The parties jointly hire a neutral third party to assist in the process. The primary difference between adjudication forums and mediation is that of decision. In adjudication (trial or arbitration), a third party finds fault and awards accordingly. In mediation, the parties define the resolution.

The mediator's role is central to the process. It is the mediator's function to hear both sides and without finding fault, assist in structuring a resolution. "Interests" rather than "positions" are defined. The successful mediator is an accomplished negotiator who attempts to help the parties choose an immediate and certain reconciliation in lieu of a distant and uncertain adjudication outcome.

The forensic engineer can play an important role in mediation. Settlement of construction cases often includes concerns regarding the cost of the repair. The forensic engineer, without attributing blame, can analyze the problem, help both sides to understand what happened and focus the discussion on why it happened. Assuming it is within the scope of the forensic engineer's engagement and area of expertise, the forensic engineer can also give parties a preview of what the positional debate at trial will look and sound like when the question of fault is explored.

Often, the issues of a construction case are clear and the culpability can be easily established. The problem simply boils down to money. That is, the defendants recognize their liability but are without the resources to respond accordingly. Mediation may offer the prospect of devising a repair that the defendants can reasonably pay for. The forensic engineer can aid this effort by identifying what is absolutely essential, what is necessary, and what could conceivably wait.

The forensic engineer called upon to participate in mediation may be of help in advising the parties where it is safe and appropriate to make concessions. To provide that input, however, the forensic engineer must be prepared to consider and help the parties fairly evaluate a full range of alternatives including those which may contradict the forensic engineer's own opinions.

6.4. ADMISSIBILITY OF TESTIMONY BY FORENSIC ENGINEERS

Admissibility of expert testimony and qualification of expert witnesses are defined and changed as a result of case law which is cited by a court or is included in rules of evidence or procedure. This section discusses Federal Rules of Evidence (2011) 702 and 703, and three cases, *Frye* (1923), *Daubert* (1993), and *Kumho* (1999).

6.4.1. Rule 702, Testimony by Expert Witnesses

In Federal Court, the Rules of Evidence are the first thresholds any expert witness must pass before testifying, and one of the first of these rules is Federal Rule of Evidence 702 (FRE 702). This rule addresses the sufficiency of the basis of an expert's testimony and states:

> *A witness who is qualified as an expert by knowledge, skill, experience, training, or education, may testify in the form of an opinion or otherwise, if:*
>
> *a) the expert's scientific, technical, or other specialized knowledge will help the trier of fact to understand the evidence or to determine a fact in issue;*
>
> *b) the testimony is based upon sufficient facts or data;*
>
> *c) the testimony is the product of reliable principles and methods, and*
>
> *d) the witness has reliably applied the principles and methods to the facts of the case.*

6.4.2. Rule 703, Bases of an Expert's Opinion Testimony

Expert witnesses may provide opinion testimony based on three general sources: firsthand observation by the witness, presentation at trial proceedings, and data presented to the expert outside of the court setting and other than from the expert's own firsthand observation. Some of that third source may be in itself inadmissible in court. Federal Rule of Evidence 703 states:

> *An expert may base an opinion on facts or data in the case that the expert has been made aware of or personally observed. If experts in the particular field would reasonably rely on those kinds of facts or data in forming an opinion on the subject, they need not be admissible for the opinion to be admitted. But if the facts or data would otherwise be inadmissible, the proponent of the opinion may disclose them to the jury only if their probative value in helping the jury evaluate the opinion substantially outweighs their prejudicial effect.*

The Rule addresses the bases of opinion testimony by experts and allows the expert to form an opinion by reasonably relying on inadmissible information, facts, and data gathered by other practitioners in the field, but does not then allow that information to be admitted simply because the expert reasonably relied upon it. The information may be disclosed to the jury, however, only for the purpose of assisting the jury to evaluate the expert's opinion.

6.4.3. *Frye*

In 1923, the Court of Appeals for the District of Columbia affirmed a lower court's decision, holding that, in order for expert testimony to be admissible, "the thing from which the deduction is made must be sufficiently established to have gained general acceptance in the particular field in which it belongs." *Frye* was a criminal case in which the defendant sought to introduce expert testimony concerning the results of a polygraph examination. The court held that the scientific community did not generally accept the polygraph examination. As a result, the polygraph examination did not rise to the level of information upon which an expert could render an opinion.

Although inconspicuous in its beginning, as the Court cited no authority to support its decision, *Frye* is an example of common law springing into reality. *Frye* became the common law relied upon by the Federal Courts concerning whether specific information was a proper basis for expert testimony. Seeking to control the scientific communities of which their products were a part, some parties held fast to the Frye doctrine in order to keep plaintiffs' legitimate evidence out of court.

Even if a party's expert was highly qualified, under Frye the party seeking to introduce the expert testimony had to prove that the information the expert relied upon in formulating his or her opinion was something collectively agreed upon by the scientific community. The Frye test became the threshold requirement for the admissibility of expert testimony in federal courts. Frye is no longer the standard used by the District Courts to evaluate opinion testimony, but does remain the law in many states, including, for instance, Illinois.

6.4.4. *Daubert*

In *Daubert*, two mothers alleged that the drug Bendectin, which they ingested while pregnant, caused limb reduction birth defects in their children. The defense presented an affidavit of their expert purporting that use of Bendectin was not causally connected to human birth defects. In response, the plaintiffs presented eight affidavits from experts delineating the causal connection between the ingestion of the drug and birth defects. Applying the *Frye* test, the trial judge found that even though the plaintiffs' experts may be highly knowledgeable, the subject matter of their testimony was not generally accepted in the scientific community. The Ninth Circuit Court of Appeals affirmed. Vacating and remanding the case, the United States Supreme Court held that the Federal Rules of Evidence superseded the *Frye* "general acceptance" test. The court reasoned that "nothing in the test of this Rule establishes 'general acceptance' as an absolute prerequisite to admissibility."

After answering the very narrow question at issue, the Supreme Court continued to analyze Rule 702 in a somewhat advisory fashion and developed a two-prong test that must be met for admitting expert scientific testimony in a federal trial court. The first prong requires that the expert testimony be based on scientific knowledge, while the second prong mandates that the testimony help the trier of fact in understanding the evidence or determining a fact in issue.

Next, to further assist the trial judge, the Court set forth four nonexclusive factors that federal judges ought to consider in carrying out their "gate keeper" function under Rule 702; 1) whether the theory or technique has been tested; 2) whether the theory or technique has been subjected to peer review and publication; 3) the known or potential rate of error, and the existence and maintenance of standards that control the technique's operation; and 4) the "general acceptance" of the scientific theory.

The above factors are merely suggestions of the Supreme Court for consideration by the federal trial judge to test the proffered evidence. In going beyond the issue presented, the Supreme Court eliminated the "general acceptance" test employed in *Frye*, making it difficult for attorneys to bring cutting-edge science into the courtroom, and, by establishing the *Daubert* criteria, required the expert witness to provide far more substantiation of his opinions than under *Frye.* Some District Courts and Appellate Circuit Courts did not believe that the Supreme Court's suggested factors for admissibility were sufficient and added an additional factor: whether the expert testimony is based on research conducted independently of litigation or expressly for the purpose of litigation. This factor poses a potential problem for the plaintiff in that the testimony is considered suspect if the research was conducted in anticipation of litigation, which is the reason the retained expert was hired—because the plaintiff intends to sue the defendant. So, the expert must defend the decision to commission testing or research, regardless of the methodology of the investigator, and it places an additional burden on the forensic engineer (in addition to the other *Daubert* tests) to maintain objectivity, not pursue research with a desired outcome in mind, and keeping an open mind and "letting the chips fall where they may."

Under Rule 702, the trial judge serves as the "gate keeper" charged with the duty of warranting that the expert's testimony "both rests on a reliable foundation and is relevant to the task at hand." In short, after the trial judge qualifies a witness as an expert in a particular field, the *Daubert* case instructs the trial judge to determine whether the expert's testimony is reliable.

Federal Rules of Evidence only apply in the Federal Courts. The *Daubert* pre-trial hearing to evaluate a potential expert witness' testimony only applies in the Federal Court. Every state has its own Rules of Evidence. Many follow the Federal model closely. The judge is always the "gate keeper" regarding the evidence going to the jury, consistent with the state's rules. Many State Courts adopted the Daubert methodology to test the value of a potential opinion witness' testimony, but not all. Some states remain "Frye" states.

6.4.5 Kumho

In March 1999, the US Supreme Court, in *Kumho Tire Co. Ltd., et al., versus Patrick Charmichael, et al.,* settled the issue of whether *Daubert* applies to engineers. The court ruled that judges can apply the same four tests, previously targeted at medical and scientific professionals, to expert testimony by engineers. A close reading of the decision however, suggests the District Court Judge recognized the inherent differences between opinion testimony advanced by a medical or scientific professional, and technical opinion testimony offered by an engineer. The Court said that not all the tests need to be applied to technical opinion testimony, and not as rigorously. Mr. Justice Breyer, writing for the majority, essentially said that the District Court Judge knows the difference between the two and can be trusted to make the proper distinction. Nevertheless, Kumho Tire made clear that an engineer's testimony may be tested by a pre-trial hearing and may be deemed inadmissible if the methods used as bases for the testimony differ from standards of practice or is not consistent with generally accepted methods.

6.5. CONCLUSION

This chapter earlier suggested that the legal forum is the domain of the lawyer, that the forensic engineer is merely litigation support, a witness whose sole task is to inform the judge and jury as to the engineering or scientific basis for a disputed fact so they can reach a verdict. Is that entirely correct? The preceding sections demonstrate forensic engineers may play far more expansive roles. They can assist in resolving the dispute without the parties bringing their case into the legal forum. In the legal forum, they can offer opinion testimony based on their own investigation. They can offer opinion testimony based upon facts and data that would be otherwise inadmissible if the value of those facts and data outweigh their possible prejudicial effect on the jury. They can offer opinion testimony that embraces an ultimate issue to be decided by the jury. They can help inform counsel and assist in shaping the litigant's case.

Many variables exist within the litigation system and within the actual process of the prosecution or defense of a lawsuit, so many that the testifying forensic engineering expert witness is unlikely to "win" or "lose" the case for the attorney and the attorney's client. But clearly the role of a testifying forensic engineer witness is extremely important, perhaps vital. The practice of forensic engineering, the forensic engineer's preparation, focus, and adherence to the ethical and professional standards advocated throughout this text must be consistent with that important role.

6.6. REFERENCES

Daubert v. Merrill Dow Pharmaceuticals, Inc., 509 U. S. 579, 113 S. Ct. 2786, 125 L. Ed. 2d 469 (1993).

Kumho Tire Co., Ltd. v. Carmichael, 131 F.3d 1433 (1999).

Federal Rules of Civil Procedure (FRCP), U.S. Government Printing Office, Washington: 2011.

Federal Rules of Evidence (FRE), U.S. Government Printing Office, Washington : 2011.

Frye v. United States, 54 App. D. C. 46, 293 F.1013 (1923).

Chapter 7
The Business of Forensic Engineering

7.1. INTRODUCTION

Starting and growing a forensic engineering business can be an exciting and challenging endeavor. Defining a plan with objectives, a strategy, and a targeted approach can lead to a steady workflow and the success of the forensic engineer and firm.

This chapter reviews some business essentials. It starts with a description of readiness for being in the business of forensic engineering. It then explores various types and modes of marketing, describes the general types of forensic work a forensic firm might accept, and touches on how and whether a forensic engineer might accept a new assignment. The chapter addresses agreements for services and types of insurance appropriate to be maintained by a forensic engineering business.

7.2. READINESS

Whether starting up a business for the first time or trying to increase an existing client base, the forensic engineer must be ready to grow the business. This takes preparation. Starting a fee-for-service business can be very challenging. The forensic engineer provides expert and unbiased opinions and therefore must be able to demonstrate the ability to carry out investigations and to report upon the findings. Qualifications for taking on a case or assignment such as education, experience, publications, and proper state licensure are essential.

The engineer should verify if holding a license to practice engineering or being registered as an engineer is required in each state where the engineer practices. Besides getting individually licensed and registered, the engineer should register the firm with the board and Secretary of State as required in the states where the practice takes place.

Initially, the forensic engineer who has interest in starting up a new business or growing the business should prepare a business plan. The business plan should outline the financial needs for the business and the first few months of operations with an economic analysis to determine the strength of the current business environment. The firm should also compile a list of competitive advantages the firm will have over others. The more the firm knows about the market, the better it can tailor the approach and service offerings. This is the time to define the target market and customer profile, determine the service pricing and project sales. Budgets and financial statements can be included. Is one forensic engineer going to run the business or is a team of forensic engineers going to be involved? What steps will be taken to initiate new business and how will the engineer respond when new business comes in? Will the forensic engineer need just an office space or an additional area such as a work lab or storage facility to use during investigations? Small Business Associations and lending institutions can provide assistance with what is needed to prepare and write a business plan.

7.3. MARKETING AND TYPES OF MARKETING

The marketing of forensic engineering services is a critical element in the effort to gain new customers and to retain clients. The goal of marketing is to get the forensic engineer's or firm's name in front of a target audience so that when the opportunity arises, a potential customer will contact the forensic engineer or firm for an assignment.

A detailed marketing strategy and marketing plan can be outlined starting with client type, work type, desired scope, and what the firm's capabilities are. The following should be taken into consideration before the firm starts to market the business: Who is the target audience? Where is the audience located? How will the firm reach the target audience, at what cost, and within what budget? Time should be taken to research which other companies, or the firm's competitors, will be within the chosen market, how the competitor reaches their target audience, and at what cost. The plan can also include different types of marketing that the firm may want to utilize.

7.3.1. Direct Marketing

Direct marketing communicates directly to the potential customer. It is one of the most important and common forms of marketing. A firm can produce, mail, and hand deliver marketing literature that describes the firm's services and contact information. Such literature can be informational brochures, advertising slicks, direct mail pieces, sample reports, resumes, postcards, and more. A cover letter and business card can be included to help introduce the forensic engineer or firm. Direct marketing also includes in-person contact, networking events, meetings, and telephone calls to reach the appropriate person who may consider hiring the forensic engineer.

7.3.2. Advertising

Firms may choose to advertise in print (newspapers, specialty magazines, publications, and journals). Specialty journals can be a good place to market to target a specific type of client. Advertising in a publication for attorneys specializing in a specific practice area can be beneficial in attaining a targeted client. Publishing special interest articles in various magazines can showcase the firm's capabilities. Mass media advertising such as radio, television, and billboards, while broadcasting to a large audience, can be expensive and not very cost effective in reaching the specific desired target audience.

There are other ways to get the firm name recognized such as free or paid advertising appearing in industry publications (insurance claims adjuster journals or websites, professional society publications, legal magazines or journals, contractor magazines, or others).

ASCE members are limited by ethical considerations in the advertising statements they can make and should be careful in crafting their media messages and public image. Advertising should be professional, in good taste, and in keeping with the Canons of ASCE. A firm looking to create a new advertisement should give it prior creative thought. The message needs to be simple and clear. State the name of the firm and what the firm does. The firm "investigates construction problems, construction accidents, building failures, building system failures, construction scheduling problems, foundation problems, soil and groundwater problems," are just a few examples. Photographs, illustrations, or logos help the audience understand in a visual way what services the firm provides.

Care should be taken to tailor the advertising message to the target market. Does the firm specialize in legal work or prefer to work directly for owners or facilities managers to identify the causes of performance problems or failures and propose solutions to their problems? Does the firm prefer to work for insurance companies helping to identify the cause of a loss and scope of repair? Whether a failure is due to an accident, the causes of the failure, and allocation of apparent responsibility are issues to be addressed by the forensic engineer.

7.3.3. Email, Websites, and the Internet

A growing form of marketing and advertising in recent years has been email. Potential clients can subscribe to email distribution lists and receive updates on various topics. The firm may choose to distribute electronic newsletters highlighting important information and updates about the firm's capabilities, their experience, and firm announcements. The frequency of such emails should be predetermined so that the sender's email is not classified as spam or so that the recipient does not feel overburdened with communication and unsubscribe from email contact from the firm.

Maintaining a professional website is important for the visibility of the firm to the public and to current clients. By providing relevant content on the website, the firm can ensure that it is promoting its areas of expertise and its professionals and their capabilities and experience. A potential client or existing client may visit the website frequently if new or updated information is provided periodically. When designing a website, it is important to know that certain types of applications on a website may make it look good, but electronic security, such as firewalls, can create difficulties for some users. For example, some Flash content on the web is not searchable. The website can also be a place where a potential client can contact the firm directly to initiate a new assignment or business need. Some firm websites have incorporated form-fill data sections where a potential client or repeat customer can enter the information about their need and how to contact them about the opportunity.

7.3.4. Directories

There are directories available for advertising opportunities, such as ALM Experts, Best's, Martindale-Hubbell, Lexis Nexis, SEAK (Skills, Education, Achievement, Knowledge), FEWA (Forensic Expert Witness Association), National Law Journal's Directory of Expert Witness and Consultants, and several others, all available on the Internet. Participating in clearing house lists also provides opportunities for forensic engineers for a fee. Others are books or CDs listed by specialty. Care should be taken, as some of these advertising services are brokers who charge a surcharge on the forensic engineer's billable rate. While brokerage services like these can serve as a good way to make client introductions, they can also make the forensic engineer's bills look inflated without adding value to the service provided.

7.3.5. Press Releases and Publications

Press releases, technical papers, and professional announcements can provide the target audience the latest information, findings, or news about a specific topic. These specialized documents educate the client and establish the forensic engineer as an authority on the subject. It is important to follow the confidentiality guidelines in place when writing about or discussing cases and assignments. It may be a violation of confidentiality for the forensic engineer to publish a paper on the topic of the investigation while the assignment is ongoing. Exceptions should be made if public safety is at issue. Clients should be informed and a copy of the paper should be shared with the client if a specific case is the topic of the paper. Technical papers can be submitted to the ASCE Forensic Congress, Structures Congress, Architectural Engineering Conferences, Journal of Performance of Constructed Facilities, or other local, national, and international professional conferences for acceptance for presentation and publication.

7.3.6. Panel Participation and Presentations

A forensic engineer may volunteer or be recruited for a formal speaking engagement. Participating as the main presenter, moderator, or panel participant can help promote the forensic engineer and firm's services. Seminars are offered by a variety of groups covering legal, engineering, financial, and insurance industry topics. Professional groups include the American Bar Association and various claims associations around the country, among others. Sample topics include insurance claims situations, construction issues, and legal case studies, oftentimes featuring photographs and diagrams. The goal of such events is to educate the audience on the topic being presented.

It can be very satisfying when the presenter is called upon for future presentations, consultations, and business opportunities. Some presentations are organized by universities or other organizations to offer continuing education to attorneys, insurance adjusters, and engineers. Participating in presentations offered by a university or a neutral third party eliminates the time, effort, and costs of marketing the seminar, getting adequate registrations of the presentations with various regulatory agencies, rental of the auditorium, and managing the event. The forensic engineer's obligation would be to prepare and present on a topic relevant to the target audience. The presenting forensic engineer may not be directly compensated for the presentation but benefits through added visibility, credibility, future business opportunities, and referrals.

Another effective marketing strategy is to offer formal continuing education courses in a classroom setting or through a webinar using the internet. A forensic engineer or firm must obtain regulatory approval to be a continuing education provider. The course outline, course description, and hours also require approval prior to offering a course.

7.3.7. Trade Show Exhibits and Conference Participation

As a means to promote a firm's expertise, the firm may choose to participate in a national or local trade show or conference session by sponsoring an event, a breakfast, lunch, or dinner break, or by exhibiting with a visual display at a booth at these types of events. This is a good opportunity to hand out firm brochures and give the potential client an opportunity to meet and visit with firm representatives.

7.3.8. Referrals

Asking for a referral from a satisfied client can be very rewarding. If the client is happy with the forensic engineer's work result, perhaps the client may be willing to pass the forensic engineer's name around or make introductions to other colleagues and acquaintances. The client and its other office locations may not be fully aware of all the capabilities and services the forensic engineer or firm can offer. If submitting a proposal to a client, it would be worthwhile to ask former clients if they would be willing to serve as references. Referrals are often the quickest, easiest, and most cost effective way to get future new business for the forensic engineer.

7.3.9. Repeat Business

Repeat business is one of the easiest ways of getting additional work. Just like referrals, repeat work is virtually a no-cost way of growing the business. By providing exceptional service to the client, the forensic engineer can market for the next project or assignment. The forensic engineer should always ask if the client's needs were satisfied by the forensic services provided, and if there is anything further the forensic engineer may do to help. If the clients are happy with the work and reports provided, it is important to make the clients aware of the other services available to them, and ask to be considered for future assignments. The forensic engineer should be sensitive in determining when the appropriate time is to ask for more work or for the next assignment. Understanding the clients and their organizations' needs is critical to maintaining a long lasting relationship.

7.4. TYPES OF ASSIGNMENTS

The forensic firm may work for legal, insurance, government, and private industry to assist with damage and condition assessments, claims disputes, subrogation efforts, and expert witness testimony.

The firm may assist plaintiffs and defense in suits regarding eminent domain, construction defect, design defect, personal injury, claims disputes and copyright issues. The assignments could include architectural engineering, structural engineering, civil engineering, mechanical engineering, and electrical engineering, among other disciplines of engineering.

Architectural engineering work may relate to building science or the technological aspects of the buildings, including the properties and behavior of building materials, components and assemblies. It may encompass foundation design, structural analysis, environmental system, construction management, or building operation. Architectural engineers are familiar with a wide range of topics including building, plumbing, electrical, mechanical, and other Codes intended to minimize hazards to occupants. Architectural engineering work involves architects, structural engineers, mechanical engineers, electrical engineers, as well as plumbing, waterproofing, weatherproofing, acoustical, and construction professionals.

Structural engineering work requires a structural expert and may include evaluating design deficiencies, professional negligence, construction defects, materials defects, structural deformation, or collapse. It may entail structural or foundation assessments after a storm, earthquake, or any other extreme or unanticipated incident. The kinds of structures may include commercial, residential, industrial, institutional buildings; power or natural resources production or distribution facilities; marine or port facilities; cranes or derricks; mining facilities, or non-building structures such as earth retaining structures, tanks, or towers.

Civil engineering work requires experts in design, construction, operation, and maintenance of drainage, grading, traffic, roadways, roadside design safety, work zone traffic control, storm water design and construction, water and wastewater systems, retaining wall design and construction, evaluations of sinkholes, and other geotechnical conditions.

Mechanical engineering work requires HVAC, boilers, and plumbing experts and includes equipment failures, HVAC, plumbing, water systems, sprinkler lines, boiler failures, high indoor humidity problems, negative building pressure, natural gas explosions, and other specific mechanical engineering issues such as, for example, the extent of damage to condenser coils after a hail event.

Forensic engineering services may be provided to many different types of clients, such as those listed in the following sections.

7.4.1. Design Engineers and Architects

It is important to know for whom and with whom the forensic engineer is working. The forensic engineer may be hired by a design engineer to perform a peer review of a project. An errors and omissions insurance carrier may want help evaluating whether the designer made an error and the extent of damage associated with another designer's error.

7.4.2. Contractors

Contractor errors can include delays that often carry a potential for large penalties or can result in a collapse or performance failure. The contractor is typically responsible for construction means and methods and when there are unforeseen problems, a forensic engineer may be well suited to troubleshoot the problem and propose solutions. A liability insurance carrier may want a review of the contractor's role in a delay or failure of the structure.

7.4.3. Building Owners

Building owners will want to know what happened and why, but will also likely want to know who they can pursue to help to pay for the necessary repairs. A building owner may need an investigation to learn if something or someone is directly responsible for the problem. The building owner will want to know if a professional's work failed to meet the standard of care. Oftentimes, a key question asked is, "Can the problem be fixed quickly and appropriately?"

7.4.4. Facilities or Property Managers

Facilities managers may have similar needs as the building owner has. However, facilities managers may focus more on the bottom-line performance of a building asset rather than what is in the long-term best interest of the building. Facilities managers may be more likely to look for opportunities to recover repair costs by suing the party most responsible. Often, the ultimate decision falls to the building owner(s), and the facilities managers get pushed aside (many examples come from homeowners' associations and condominium associations).

7.4.5. Attorneys

To be considered unbiased and objective, forensic engineers should make their services available to both defense and plaintiff attorneys.

Plaintiff's attorneys generally represent parties that consider themselves to have been wronged. Examples include a construction worker injured in an accident, an owner suffering ill effects of a poorly constructed building, or an insurance company that has paid a claim but now wants to subrogate against the alleged responsible party to be reimbursed for their cost to settle the loss or claim.

Defense attorneys generally defend those accused of wrongdoing or who otherwise are at risk of losing money if a claim is alleged against them. Examples include a scaffold manufacturer in the event of a scaffolding accident, the scaffold erector in the same event, the subcontractor who required the scaffold, a concrete supplier in the event of a concrete material failure, or a general contractor when a subcontractor caused a delay that result in additional time and costs to the owner. Defense attorneys may hire a forensic engineer to assess the damage to a building when a building owner is claiming that they have not been adequately compensated for the loss resulting from a covered peril like a storm event.

7.4.6. Insurance Company Adjusters and Independent Adjusters

An insurance company adjuster is one who is employed by the insurance company; the company adjuster works directly for the insurance carrier. A forensic engineer may start an assignment working with one adjuster but then work with another adjuster from the same firm, perhaps at a higher management level, if the loss is large.

By contrast, an independent adjuster may work for many different insurance carriers. Adjusters are generally qualified and chosen based upon the size of losses they are entitled to adjust. The type of assignments for independent and insurance company adjusters would typically entail assessing damage to a property after a peril like a fire or a storm event. The forensic engineer may be required to develop a scope of repair. In some instances the engineer may be hired to prepare construction drawings for repair.

7.4.7. Public Adjusters

Public adjusters work for the insured, perhaps on a contingent fee basis, and may work in conjunction with a plaintiff's attorney. Public adjusters may retain a forensic engineer to assist their side in determining the scope and effect of damage from a covered loss.

7.5. WHEN THE PHONE RINGS

The first step when evaluating an opportunity for new business, after clearing any conflicts, is to qualify the opportunity by gathering the initial information. Details should be noted and logged, such as the client name, company name, address, contact information, phone number, loss identification, loss location, loss party, structure information (foundation, framing, finishes, plans, details, roofing, geotechnical information), scope of work, and proposed timeline. Information should be collected documenting the type of loss, what it affected, the expertise needed and the description of the scope of the assignment. The forensic engineer should be prepared to ask questions and request any additional information needed to clarify the scope of the assignment if it is not clear in the initial phone call. Engineers who are prepared have a better chance of being retained for the work.

It is critical to conduct a conflict of interest check within the firm prior to accepting the opportunity. The firm files and database should be reviewed for owner names, street names, specific addresses, attorney contacts, and insurance company for information that may cause a conflict prior to accepting work. In a large firm, a process should be established to check for conflicts with all branches of offices.

7.6. WHAT TO TAKE AND WHAT TO PASS

Before accepting the assignment, the forensic engineer should verify that the firm is capable of doing the assignment and that equipment or supplies for the assignment are available and the location for the assignment is feasible within the projected timeline of service.

From time to time, a firm may have to decline working on an assignment. It may be because the firm does not have the licensure or the qualified engineers to do the work. Perhaps the resources to complete the assignment are not available. The forensic engineer may be unable to conduct all the research necessary. A conflict of interest may be present or may arise. A time constraint may be present. The forensic engineer may have a budget that needs to be met as defined by the client. It is best to have fee discussions up front with the client before starting the work. The forensic engineer should relay initial fee information and answer any questions, perhaps even asking if there is a cap on fees, before beginning the work. A client may have unrealistic limits on a maximum fee or may be unreasonable in not reimbursing travel and other out of pocket costs. These are all valid reasons for declining work.

7.7. AFTER THE DECISION TO TAKE

When taking on an assignment, there are a few things that need to be determined at the start. The forensic engineer should understand the client expectations. Will a retainer be required before starting the work for the client?

It is typical to have a retainer on file for a new client? Some companies may choose to charge a retainer on a case-by-case basis. The agreement should define how often the client will be billed for services.

The type of report to be provided should be defined, For example, will the client require a verbal report, a short written letter-report, or a comprehensive written report? Does the forensic engineer have a checklist in place that details what will be included in the report, such as narrative, photographs, illustrations, and appendices? A report format should be established for completed assignments and investigations.

What is the estimated timeline from the initial site visit and investigation to the delivery of the report? What additional items are required to be in the report? Are there any special client requests?

The forensic engineer or firm should have a file numbering system and tracking system in place. At minimum, prepare for clients to inquire about the status of an assignment, and how that information can be logged, tracked, and queried easily. Some may choose to use an electronic database and computer software for tracking files and assignments.

7.8. LETTER OF AGREEMENT

An agreement letter, engagement letter or proposal contract should be sent to the client and signed before the forensic engineer begins any work. It is a written scope of the professional services—what services will be provided and by whom. It should include the fees and any additional charges for professional services, the terms of the agreement, including a retainer request, and can include any other important information. The agreement or contract will most likely be different for insurance clients and for legal or private clients. An insurance company may provide the forensic engineer with a Statement of Work or a short agreement that includes professional fees and charges on a time and materials basis.

The forensic engineer should exercise extreme caution in accepting any form of agreement provided by a potential client, and should have such agreements carefully reviewed by counsel and, if warranted, by the forensic engineer's professional liability insurance carrier.

7.8.1. Fee Structure or Pricing

The most common arrangement is an hourly fee for services. The forensic engineer keeps track of time spent on various tasks and bills on the basis of hours and fractions of hours,. It is common to charge different rates for different personnel. For example, junior engineers may perform library research to track down and copy relevant resource materials. The principal of a firm may have a higher hourly rate than that of a project engineer.

In the event that the services of a subcontractor or sub-consultant are engaged by the forensic engineer, the forensic engineer needs to decide how those services will be billed to the client. For example, a contractor accompanies an investigator into the field to cut wall openings that permit inspection and then repairs them after the forensic engineer has seen the exposed conditions. Another example may be a specialty testing lab that performs rigorous tests to ASTM standards for a lump sum, or a specialty firm that performs a complex computer analysis of structural systems to support the forensic engineer's work. When billing for a subcontractor or sub-consultant, if the client agrees, it is acceptable to mark up the subcontractor or sub-consultant invoice to account for administrative fees and for the cost of money. (The forensic engineer should not serve as a lending bank for the client.)

It is good business practice, and assures continued availability and responsiveness of subcontractors and sub-consultants, to pay their invoices in a timely fashion independent of payment by the client. For maintaining good business relationships, the forensic engineer's agreement with the subcontractor or sub-consultant should not be on a basis of "pay-if-paid." It is also appropriate to impose late and service charges and interest on invoiced amounts not in dispute that have not been paid in accordance with the payment schedule that was agreed upon. The disadvantage of retaining subcontractors or sub-consultants is that the forensic engineer or firm may then be assumed to be responsible for the work or services of those parties.

A fixed fee structure may be employed for phased or clearly defined services. An example of a fixed fee is when the forensic engineer knows very well what the scope of services will cost and has control over them. This is when a fixed fee may be appropriate. Fixed fee arrangements are often preferred by clients because of their certainty.

Forensic engineers may set unit fees for specific tasks. They may provide an estimate based on an expected number of days in the field and also have a cost-per-day allowance for extra days. In a situation where documents must be reviewed, it may help the client if there is an agreement in place regarding the costs associated with reading depositions or reviewing documents.

When the overall scope is uncertain, it may make sense to phase the work, carefully outlining an estimate for the first phase, but recognizing that the requirements of subsequent phases will depend upon the outcome of that first phase.

The firm may need to have alternative fee structures for emergency services. For example, after a collapse or catastrophe, there may be no time to prepare a careful proposal, scope of work and fee estimate. The forensic engineer may be required to:

- Evaluate extent of damage.
- Design immediate temporary access to allow investigators safe access.
- Establish protocols for several different investigators to access the site.
- Establish protocols for handling and storing physical evidence.
- Design measures for physically and functionally isolating the site.
- Design temporary repairs.
- Monitor the temporary construction.
- Prepare and perform the investigation.
- Prepare repair schemes and procedures.

Regardless of the pricing scheme, fees should not be tied to an outcome. Forensic engineers must not have financial incentives attached to their unbiased findings.

Fees must also reflect costs. The expenses that the forensic engineer incurs need to be considered. Travel and other out-of-pocket expenses are usually extra, and many include a handling fee determined by the forensic engineer or firm. Other expenses vary widely. Others may include computer time within the forensic engineer's firm, camera work, photograph reproduction, specialty equipment charge, and storage for specimens and samples. Some of these may legitimately be considered overhead and be included in other costs related to the business such as insurance, benefits, holidays, paid time off, and the cost of carrying payroll between the time checks are written and when monies are collected on the invoices.

Overall, the biggest concern should be whether the firm is making a reasonable profit with the pricing structure. Company expenses, operations, and overhead costs must be paid for out of the money earned working directly for a client on billable work. If a forensic engineer needs to charge more than the best competitor to keep the business going, then the forensic engineer's reputation or level of service must outweigh that competitor's reputation or level of service. If it does not, then the forensic engineer should identify operational inefficiencies or opportunities for cost control within the operation.

7.9. LIABILITY

Engineering professionals develop opinions upon which the client will rely. They provide many services that can include property assessments, structural investigations, pre-purchase condition assessments, construction administration, environmental assessments, bridge inspections, hydraulic analyses, traffic impact analysis, and forensic engineering and investigations.

Several areas of potential liability exist within forensic engineering including loss of or damage to material evidence, perceived or real conflicts of interest, errors in documentation, errors in data collection and errors in the analysis of data. These are generally considered pre-trial litigation support services. In most cases, a third party is affected by the opinions developed by consultants providing forensic engineering services.

Failure to meet the standard of care of a professional engineer in performing an investigation can lead to problems for the client as well as the forensic engineer.

7.9.1 Professional Liability Policies

There are many aspects of professional liability insurance policies that should be considered when evaluating policies, including:

- Limits of the policy per claim and per year.
- Amounts of occurrence and aggregate deductibles.
- Availability of coverage for prior acts.
- Aspects of the professional practice that are or are not covered.
- How legal expenses are applied against the coverage limit.
- What impact on coverage or exposure refusal to accept a settlement offer reached between the claimant and the carrier might have.
- Who selects counsel.
- What affect claims or the lack thereof have on premiums.

7.10. INSURANCE

From the insured's perspective, an insurance policy is written to provide coverage for and protect the financial well being of the insured. Insurance policies are available to cover almost every aspect of our lives including our property, our actions, and even our future. The cost of insurance policies is driven by a number of factors including the amount of coverage provided, the risk being insured against, the history of risk associated with the coverage, the insured's history of losses, and the normal pressures of the marketplace.

Insurance is an integral aspect of the forensic engineer's daily practice. It can affect the type of services offered and to whom those services are provided. It may limit the engineer's potential clients.

In some cases, fees generated by a firm for a service may not justify the cost of the insurance premium. In other cases, clients may require the engineer to maintain levels of insurance in excess of what the engineer normally carries, attempt to incorporate contract language that nullifies coverage, or require the engineer to obtain coverage that in fact is not available in the market. Forensic engineers should be mindful of insurance requirements imposed by potential clients.

7.10.1. Types of Insurance

No one insurance policy is sufficient to cover all aspects of an engineering practice. Most businesses provide benefits in the form of workers' compensation, health, life, and disability insurance for employees. Other policies may be required under different circumstances. Most forms of insurance fall under one of the following categories.

7.10.1.1. Auto Liability

Auto liability insurance provides coverage for losses caused by injuries or damage to property and legal liability imposed on the insured for such injury or for damage to property caused in conjunction with operation of a vehicle.

7.10.1.2. Builder's Risk

Builder's risk policies are normally carried by a contractor or owner. Also termed All Risk or Course of Construction insurance, it is intended to cover both the labor and materials necessary to rebuild in the event of damage or destruction during the course of construction.

7.10.1.3. General Liability

General liability insurance provides coverage for legal liability to third parties that are not a result of professionally negligent acts, errors, or omissions. General liability insurance commonly applies to exposure created by the premises and a firm's operations.

7.10.1.4. Non-Owned Auto Insurance

Non-owned auto insurance provides coverage for liability and property damage claims which may be caused by automobiles not owned or hired by the firm.

7.10.1.5. Professional Liability Insurance

Also called errors and omissions insurance, professional liability insurance protects the professional against claims arising from negligent acts or omissions in the performance of professional services.

7.10.1.6. Project Insurance

Project insurance provides project-specific professional liability coverage for one or all members of a design or construction team on a given project.

7.10.1.7. Valuable Papers Coverage

Valuable papers insurance covers some of the cost of replacing documents lost due to fire, water damage or other specified perils. It usually is limited to a particular amount, typically less than the actual cost of recreating all the lost documents from scratch.

7.10.1.8. Workers' Compensation

All employers are typically required by state law to purchase workers' compensation insurance. It provides predetermined benefits for deaths or injuries caused by the conditions of the work place.

7.11. INDIVIDUAL VS. GROUP PRACTICE

Expanding from an individual practice to a group practice may broaden a forensic engineer's ability to serve clients. The client may look to one source for resources relating to the assignment. On the other hand, there may be significant advantages for a practitioner to maintain an individual practice.

7.11.1. Group Practice

Most major and large-scale investigations require assembling a team of diversified forensic professionals who have different areas of expertise for the forensic investigation. The multidisciplinary firm can offer the potential client expertise in several areas. For example, a client may need an architect, a civil engineer, a structural and geotechnical engineer, a materials engineer, a mechanical engineer, an electrical engineer and a cost estimator. A multidisciplinary firm may be able to service the needs together from one firm. The firm may appoint a lead forensic investigator who serves as team leader and also as the lead or testifying expert witness. Not all firms employ individual experts in each required specialty. Sometimes, even very large firms may have to bring in outside resources.

There may be cost advantages resulting from **expansion into** a group practice or multidisciplinary firm. Professionals working at the same location can share office space, computer and office equipment, telephone, and software systems. Administrative support staff can assist the forensic engineering group with daily operational tasks or administrative duties. Assignment intake, invoicing, and other overhead functions can be centralized.

The multidisciplinary firm or group practice can create separate administrative and management departments to monitor and comply with insurance requirements and state licensing, and to plan the logistics of serving the client, possibly allowing the forensic engineers of the firm to concentrate on investigation, reporting and testifying.

Advantages for operating a multidisciplinary forensic engineering firm might include:

- Scheduling of professionals.
- Economies of scale.
- Multiple areas of expertise.
- "One stop" service provider.
- Overhead savings.
- Establishing a strong, agile corporate image.

One may effectively argue that many of these characteristics of larger firms also exist in small firms or individual practices, and in some cases may be more descriptive of smaller firms.

7.11.2 Individual Practice

A large number, if not most, claims that need expert consultation and witnessing are such in size and specialty that an individual practitioner can handle them by himself or with minimal outside support.

When major and large-scale investigations require the effort of more than one individual, or the assembling a team of forensic professionals who have different areas of expertise, a large firm is not always a necessity; that team may be assembled and managed by a sole practitioner who retains sub-consultants in the needed disciplines and fields. (See section 2.1.4. Generalist Forensic Engineer, above.)

Many forensic assignments do not require an extensive team of experts. In either situation, there are advantages of and for the individual practitioners. Robert T. Ratay, an author and contributor to these *Guidelines* among many other publications, wrote in Structure Magazine (2007):

> *While most of the forensic structural engineering work is performed by engineering firms of various sizes, a fair amount is performed by individual practitioners. Individuals with long-time design and construction experience [may] ... step into the practice; university professors make themselves available when the opportunity arises.*
>
> *An engineer should seriously consider all of the pros and cons before deciding to enter into private practice. It is challenging, satisfying and [potentially] lucrative work but the world of adversity is not always pleasant. A client in need of expert engineering advice and litigation support should weigh all the pros and cons in light of the manpower needed for the project; decide whether [the] need is for only one particular specialty or for a range of expertise; and consider the level of trust and comfort with the individuals who will be doing the actual technical work, writing the reports, giving depositions, participating in dispute resolutions and testifying in court in the client's behalf.*

7.12. CONCLUSION

The business of forensic engineering encompasses the readiness of a forensic engineer and qualifications to start and operate a business. Strategic planning and different types of marketing and advertising efforts can help in getting the forensic engineer's or firm's name out to the target audience. Honesty and sincerity to its customers will take the firm to higher levels, allowing the firm to branch out over time. Ultimately, the approach, methodology, fee structure, and quality of service will result in a satisfied customer base and earn the firm referrals and repeat business for a long-lived, successful, and growing business practice.

Chapter 8
Summary

The purpose of these *Guidelines for Forensic Engineering Practice* is to report the Forensic Practices Committee's current understanding of the state of acceptable forensic civil engineering practice in the United States. The impetus for its publication arose from the realization that the practice and understanding of forensic engineering has changed since the first edition of the *Guidelines for Forensic Engineering Practice* was published in 2003.

This Second Edition of the *Guidelines* updates and expands on topics concerning forensic engineers presented in the first edition in 2003. The introductory chapter defines forensics and forensic engineering, and presents abstracts of the subsequent chapters. Those include a new chapter on the standard of care, and revised and updated chapters on the competencies and qualifications of forensic engineers, investigations and reports, ethics, the legal forum, and the business of forensic engineering.

The chapter on competencies and qualifications of forensic engineers described skill sets appropriate for a practicing forensic engineer and clarified the fact that the forensic engineer is qualified as an expert only by the court. "Forensic engineering" is not yet a licensed or registered engineering practice or discipline, and neither the States nor the non-profit professional societies qualify forensic engineers.

The chapter on the standard of care characterizes that term as the boundary between negligent and non-negligent error. The standard of care can be thought of as the duty of the engineer to exercise care, and is only determined when a question arises concerning an engineer's negligence. The chapter also acknowledges that the determination of the standard of care and a practitioner's fulfillment of that duty are questions of fact that a judge or jury can answer. To help the judge or jury answer those questions, expert witnesses may give opinion testimony developed using valid and reliable methods. The chapter includes examples of both invalid and unreliable methods that were used as a basis of expert testimony and a proposed valid, reliable method.

The chapter on investigation and reports describes the ingredients and features of a forensic engineering investigation and report. The Committee on Forensic Investigations of the TCFE is in the process of writing and publishing their *Guidelines for Failure Investigations*. The reader is encouraged to refer to that publication for more information.

The chapter on ethics describes how the principles and canons of the ASCE Code of Ethics are integral to every aspect of forensic engineering practice. As an example of the application and understanding of ethics, the chapter presents a case study of the investigations and opinions of forensic engineers on two sides of an issue.

Most other professional societies have Codes of Ethics that their members are expected to follow. The reader is encouraged to obtain information about the Codes of Ethics of other organizations that may be relevant.

The chapter on the legal forum describes the legal environment in which forensic engineers may work and many of the procedures and rules that expert witnesses follow and by which they are governed.

The chapter on the business of forensic engineering outlines many of the business considerations that affect a forensic practitioner or firm.

These *Guidelines* are a continuation of the FPC's own learning and understanding of the field and practice of forensic engineering. The FPC encourages the reader to continue learning about the field and wants these *Guidelines* to serve as a springboard for more in-depth learning, and for the advancement of the practice of forensic engineering.

Appendix
American Society of Civil Engineers
Code of Ethics[1]

Fundamental Principles[2]

Engineers uphold and advance the integrity, honor and dignity of the engineering profession by:

1. using their knowledge and skill for the enhancement of human welfare and the environment;
2. being honest and impartial and serving with fidelity the public, their employers and clients;
3. striving to increase the competence and prestige of the engineering profession; and
4. supporting the professional and technical societies of their disciplines.

Fundamental Canons

1. Engineers shall hold paramount the safety, health and welfare of the public and shall strive to comply with the principles of sustainable development[3] in the performance of their professional duties.
2. Engineers shall perform services only in areas of their competence.
3. Engineers shall issue public statements only in an objective and truthful manner.
4. Engineers shall act in professional matters for each employer or client as faithful agents or trustees, and shall avoid conflicts of interest.
5. Engineers shall build their professional reputation on the merit of their services and shall not compete unfairly with others.

---

[1] The Society's Code of Ethics was adopted on September 2, 1914 and was most recently amended on July 23, 2006. Pursuant to the Society's Bylaws, it is the duty of every Society member to report promptly to the Committee on Professional Conduct any observed violation of the Code of Ethics.

[2] In April 1975, the ASCE Board of Direction adopted the fundamental principles of the Code of Ethics of Engineers as accepted by the Accreditation Board for Engineering and Technology, Inc. (ABET).

[3] In October 2009, the ASCE Board of Direction adopted the following definition of Sustainable Development: "Sustainable Development is the process of applying natural, human, and economic resources to enhance the safety, welfare, and quality of life for all of the society while maintaining the availability of the remaining natural resources."

6. Engineers shall act in such a manner as to uphold and enhance the honor, integrity, and dignity of the engineering profession and shall act with zero-tolerance for bribery, fraud, and corruption.
7. Engineers shall continue their professional development throughout their careers, and shall provide opportunities for the professional development of those engineers under their supervision.

Guidelines to Practice Under the Fundamental Canons of Ethics

CANON 1.
Engineers shall hold paramount the safety, health and welfare of the public and shall strive to comply with the principles of sustainable development in the performance of their professional duties.

a. Engineers shall recognize that the lives, safety, health and welfare of the general public are dependent upon engineering judgments, decisions and practices incorporated into structures, machines, products, processes and devices.
b. Engineers shall approve or seal only those design documents, reviewed or prepared by them, which are determined to be safe for public health and welfare in conformity with accepted engineering standards.
c. Engineers whose professional judgment is overruled under circumstances where the safety, health and welfare of the public are endangered, or the principles of sustainable development ignored, shall inform their clients or employers of the possible consequences.
d. Engineers who have knowledge or reason to believe that another person or firm may be in violation of any of the provisions of Canon 1 shall present such information to the proper authority in writing and shall cooperate with the proper authority in furnishing such further information or assistance as may be required.
e. Engineers should seek opportunities to be of constructive service in civic affairs and work for the advancement of the safety, health and well-being of their communities, and the protection of the environment through the practice of sustainable development.
f. Engineers should be committed to improving the environment by adherence to the principles of sustainable development so as to enhance the quality of life of the general public.

CANON 2.
Engineers shall perform services only in areas of their competence.

a. Engineers shall undertake to perform engineering assignments only when qualified by education or experience in the technical field of engineering involved.

b. Engineers may accept an assignment requiring education or experience outside of their own fields of competence, provided their services are restricted to those phases of the project in which they are qualified. All other phases of such project shall be performed by qualified associates, consultants, or employees.
c. Engineers shall not affix their signatures or seals to any engineering plan or document dealing with subject matter in which they lack competence by virtue of education or experience or to any such plan or document not reviewed or prepared under their supervisory control.

CANON 3.
Engineers shall issue public statements only in an objective and truthful manner.

a. Engineers should endeavor to extend the public knowledge of engineering and sustainable development, and shall not participate in the dissemination of untrue, unfair or exaggerated statements regarding engineering.
b. Engineers shall be objective and truthful in professional reports, statements, or testimony. They shall include all relevant and pertinent information in such reports, statements, or testimony.
c. Engineers, when serving as expert witnesses, shall express an engineering opinion only when it is founded upon adequate knowledge of the facts, upon a background of technical competence, and upon honest conviction.
d. Engineers shall issue no statements, criticisms, or arguments on engineering matters which are inspired or paid for by interested parties, unless they indicate on whose behalf the statements are made.
e. Engineers shall be dignified and modest in explaining their work and merit, and will avoid any act tending to promote their own interests at the expense of the integrity, honor and dignity of the profession.

CANON 4.
Engineers shall act in professional matters for each employer or client as faithful agents or trustees, and shall avoid conflicts of interest.

a. Engineers shall avoid all known or potential conflicts of interest with their employers or clients and shall promptly inform their employers or clients of any business association, interests, or circumstances which could influence their judgment or the quality of their services.
b. Engineers shall not accept compensation from more than one party for services on the same project, or for services pertaining to the same project, unless the circumstances are fully disclosed to and agreed to, by all interested parties.

c. Engineers shall not solicit or accept gratuities, directly or indirectly, from contractors, their agents, or other parties dealing with their clients or employers in connection with work for which they are responsible.
d. Engineers in public service as members, advisors, or employees of a governmental body or department shall not participate in considerations or actions with respect to services solicited or provided by them or their organization in private or public engineering practice.
e. Engineers shall advise their employers or clients when, as a result of their studies, they believe a project will not be successful.
f. Engineers shall not use confidential information coming to them in the course of their assignments as a means of making personal profit if such action is adverse to the interests of their clients, employers or the public.
g. Engineers shall not accept professional employment outside of their regular work or interest without the knowledge of their employers.

CANON 5.
Engineers shall build their professional reputation on the merit of their services and shall not compete unfairly with others.

a. Engineers shall not give, solicit or receive either directly or indirectly, any political contribution, gratuity, or unlawful consideration in order to secure work, exclusive of securing salaried positions through employment agencies.
b. Engineers should negotiate contracts for professional services fairly and on the basis of demonstrated competence and qualifications for the type of professional service required.
c. Engineers may request, propose or accept professional commissions on a contingent basis only under circumstances in which their professional judgments would not be compromised.
d. Engineers shall not falsify or permit misrepresentation of their academic or professional qualifications or experience.
e. Engineers shall give proper credit for engineering work to those to whom credit is due, and shall recognize the proprietary interests of others. Whenever possible, they shall name the person or persons who may be responsible for designs, inventions, writings or other accomplishments.
f. Engineers may advertise professional services in a way that does not contain misleading language or is in any other manner derogatory to the dignity of the profession. Examples of permissible advertising are as follows:

    - Professional cards in recognized, dignified publications, and listings in rosters or directories published by responsible organizations, provided that the cards or listings are consistent in size and content and are in a section of the publication regularly devoted to such professional cards.

- Brochures which factually describe experience, facilities, personnel and capacity to render service, providing they are not misleading with respect to the engineer's participation in projects described.
- Display advertising in recognized dignified business and professional publications, providing it is factual and is not misleading with respect to the engineer's extent of participation in projects described.
- A statement of the engineers' names or the name of the firm and statement of the type of service posted on projects for which they render services.
- Preparation or authorization of descriptive articles for the lay or technical press, which are factual and dignified. Such articles shall not imply anything more than direct participation in the project described.
- Permission by engineers for their names to be used in commercial advertisements, such as may be published by contractors, material suppliers, etc., only by means of a modest, dignified notation acknowledging the engineers' participation in the project described. Such permission shall not include public endorsement of proprietary products.

g. Engineers shall not maliciously or falsely, directly or indirectly, injure the professional reputation, prospects, practice or employment of another engineer or indiscriminately criticize another's work.
h. Engineers shall not use equipment, supplies, laboratory or office facilities of their employers to carry on outside private practice without the consent of their employers.

CANON 6.
Engineers shall act in such a manner as to uphold and enhance the honor, integrity, and dignity of the engineering profession and shall act with zero tolerance for bribery, fraud, and corruption.

a. Engineers shall not knowingly engage in business or professional practices of a fraudulent, dishonest or unethical nature.
b. Engineers shall be scrupulously honest in their control and spending of monies, and promote effective use of resources through open, honest and impartial service with fidelity to the public, employers associates and clients.
c. Engineers shall act with zero-tolerance for bribery, fraud, and corruption in all engineering or construction activities in which they are engaged.
d. Engineers should be especially vigilant to maintain appropriate ethical behavior where payments of gratuities or bribes are institutionalized practices.

e. Engineers should strive for transparency in the procurement and execution of projects. Transparency includes disclosure of names, addresses, purposes, and fees or commissions paid for all agents facilitating projects.
f. Engineers should encourage the use of certifications specifying zero tolerance for bribery, fraud, and corruption in all contracts.

CANON 7.
Engineers shall continue their professional development throughout their careers, and shall provide opportunities for the professional development of those engineers under their supervision.

a. Engineers should keep current in their specialty fields by engaging in professional practice, participating in continuing education courses, reading in the technical literature, and attending professional meetings and seminars.
b. Engineers should encourage their engineering employees to become registered at the earliest possible date.
c. Engineers should encourage engineering employees to attend and present papers at professional and technical society meetings.
d. Engineers shall uphold the principle of mutually satisfying relationships between employers and employees with respect to terms of employment including professional grade descriptions, salary ranges, and fringe benefits.

Index